JAGUAR XK ENGINE

History, buying, rebuilding, uprating

Dave Pollard

Haynes®

Front cover illustrations: The background cross-section drawing of a Jaguar XK engine is courtesy Ford of Europe Inc. The cutaway illustration of a Jaguar XK120 of 1948 is taken from *Inside 100 Great Cars*, published by G. T. Foulis & Co.

Back cover illustration: An engine from a Jaguar XK150. (Courtesy Haynes Motor Museum)

First published in 1998

British Library cataloguing-in-publication data:
A catalogue record for this book is available from the British Library.

ISBN 1 85960 007 7

Library of Congress Catalog Card Number 98-70518

Haynes North America Inc.
861 Lawrence Drive, Newbury Park,
California 91320, USA

Published by Haynes Publishing, Sparkford,
Nr Yeovil, Somerset BA22 7JJ, UK.

Tel: 01963 440635 Fax: 01963 440001
Int. tel: +44 1963 440635 Fax: +44 1963 440001

E-mail: sales@haynes-manuals.co.uk
Web site: http://www.haynes.com

Printed and bound by in Great Britain by J. H. Haynes & Co. Ltd

Jurisdictions which have strict emission control laws may consider any modification to a vehicle to be an infringement of those laws. You are advised to check with the appropriate body or authority whether your proposed modification complies fully with the law. The publishers accept no liability in this regard.

While every effort is taken to ensure the accuracy of the information given in this book, no liability can be accepted by the author or publishers for any loss, damage or injury caused by errors in, or omissions from, the information given.

Contents

Acknowledgements

It is only through the kind co-operation of many companies and individuals that it has been possible to write this book. Those mentioned below (and others too many to mention here – but included in the text of the book and the appendix at the back) have not only provided technical information and photographs, but also the benefit of their many years of Jaguar experience. My thanks to them all.

Julia Simpson was invaluable in sourcing many of the archive pictures from the Jaguar-Daimler Heritage Trust.

When it comes to picking a company to handle your XK engine, you need someone who has been in the business long enough to know all the wrinkles (and there are plenty of them) and, ideally, someone who practises what he preaches. Tim Camp and Rob Beere between them represent a wealth of experience. The main engine rebuild section relied on the expertise of Tim Camp's highly-skilled team at VSE Engineering situated in the heart of rural Wales. In three days they turned an oily, dirty and very tired 4.2 litre XK engine into a gleaming, purring pussycat; offering, in the process, many technical asides and gems of information to aid the DIY engine builder. In contrast to Tim Camp's rural hideaway, Rob Beere Racing Services, specialists in creating racing engines, is located in the industrial heart of the country – Coventry, appropriately.

My thanks to all three for their input, particularly with regard to tuning the XK from a purring pussycat into a howling wild animal – and all points in-between.

Clarke International is world-renowned for its huge range of power tools, notably welding and pneumatic equipment, and my thanks go to Ross Bernard for the supply of tools, photos and information for the relevant sections.

Hand tools from Sykes-Pickavant were used for many of the sections, and I am grateful for the technical help and photos provided by Christine Whitworth.

Last … but definitely not least, my wife Ann has been instrumental in piecing everything together, checking and proofing, and generally putting up with endless discussions about UK-built 6-cylinder engines.

Introduction

It is now 50 years since the Jaguar XK engine was introduced, and for almost four decades this wonderfully smooth power unit propelled *all* Jaguar's 6-cylinder cars. Such longevity is astounding when viewed from today, when three years is a model-span and five years is positively archaic. Nevertheless, throughout its long life the XK engine was able to meet all the performance extremes that were demanded of it, from the rorty fire-breathing XK 150 3.8S and E-types to the quiet and graceful Mk VII and XJ6 saloons and, of course, the Le Mans-winning C- and D-types which scored wins on no fewer than five occasions.

As if all that weren't enough, it was also used as the basis for specialist applications as far apart as Dennis fire tenders, offshore power boats and even military tanks!

You might think that an engine in production for so many years is just further evidence of the all-too-typical head in the sand attitude of the British motor industry, but this is not so, and when it first appeared at the 1948 London Motor Show, under the bonnet of an XK 120 sports car, it was well ahead of its time. The basic premise was so 'right' and it gave the engine a life much longer than could reasonably be expected.

PART 1

Heritage

Chapter 1

History of Jaguar and the XK Engine

To consider anything Jaguar is to consider William Lyons (Sir William from 1956). Like many great pioneering entrepreneurs, his product wasn't just something to make money, it was a way of life – a life he loved. He was always involved at ground level with all his cars, often actually taking the helm in design terms. And he realised that even if you've got the greatest product in the world, it's quite another thing to get the public to buy it. History shows that William Lyons always had his finger firmly on the PR pulse.

In 1923, Lyons joined forces with a close neighbour of his who was making motorcycle sidecars in his garage. The neighbour, William Walmsley, was selling with no difficulty all he could make, and was quite happy to carry on doing so, but Lyons could see far greater potential in this cottage industry. The partnership began life as the Swallow Sidecar Company, and the expected increase in the flow of work led to rapid expansion.

As it grew, the name and operations parameters changed to include coachbuilding work, and by 1927 the company was offering a rebodied version of the popular Austin 7. Very much a working man's car, the opportunity to have what looked like a much more upmarket vehicle for little more than the cost of the original made it very popular. Indeed, his first order from a London garage was for an incredible 500 cars! That Lyons (ignoring his older partner's understandable concern) boldly accepted the order and delivery dates with absolutely no hesitation was the mark of a man destined to become one of the country's great entrepreneurs.

Continuing expansion meant that a move was required, and the most obvious place to look was in the industrial heart of the country – fittingly, the Midlands. In 1928 in Foleshill, Coventry, an ex-wartime munitions factory was geared up to make something far more useful – and profitable. The Swallow Company continued its activity of fitting new bodies to existing chassis and engine layouts and, once ensconced in its new factory, Lyons approached the Standard company with a view to producing a car based on the Standard Nine model. When Lyons wanted to produce a more sporting car, he realised that none of the Standard chassis was suitable, so a special was produced for him which, suitably clothed in curvaceous Swallow bodywork, was shown in 1931 at the London Motor Show and named the SS. There were two models – the 16hp and 20hp SS1 and the 9hp SS2. What the initials stand for has never been confirmed, although in retrospect, it would have been a masterful piece of PR hype to deliberately retain an air of mystery about them in order to create more column inches of free publicity. Regardless of what or why, the three SS models proved a huge success, their rakish lines having much to do with it, as they were not particularly impressive performers. The Mk 2 version of the SS was launched in 1934, a year not without other significance – William Walmsley left the company which became a public limited company by the name of SS Cars Limited.

The following year, William Lyons (now the sole proprietor of the business), launched the Jaguar name for a new range of cars in the pipeline, and introduced the 2-seater SS90 which was powered by a 2663cc Standard engine. However, Lyons realised that to really make an impression more power was required, and to this end he brought in Harry Weslake, already a legend in the field of cylinder head design. Weslake introduced the crossflow head design for the original Standard engine which upped the power from approximately 75bhp to a more healthy 100bhp when fitted into the SS100. When the engine capacity was increased to 3.5 litres for the 1938 model year, it was able to top 100mph – no mean feat all those years ago.

Meanwhile, Lyons was drawing around him a team of experts that would be instrumental in producing the XK engine. In 1935 Bill Heynes was taken on to head-up the Engineering Department, and three years later he was joined by Wally Hassan. Just before the outbreak of the Second World War, Lyons employed Claude Baily as Chief Designer. For five years thereafter the Foleshill factory was engaged in war work which extended the company's skills and gave it valuable experience. Immediately after the war, the company name was changed to Jaguar Cars Limited – the letters SS having acquired an unfortunate connotation by being the same used by the sinister élite corps of the by then vanquished Nazi regime in Germany.

THE XK PROJECT GATHERS MOMENTUM

The XK engine first saw the (public) light of day in 1948, clothed in the beautiful bodywork of the XK 120, but if it hadn't been for the war, it would have had an even longer heritage, as it was conceived in essence as far back as 1939. At the time, the company was powering the SS100 sports car with the over-developed 3.5-litre push-rod engine. Although this had been tuned to produce 160bhp, it had to be taken right to the limit to do so and, as ever in these cases, ultimate power was obtained at the expense of reliability and tractability.

The basic concept for the new engine came together during Lyons' stints on wartime firewatching duty along with his chief engineer, Bill Heynes, and other members of his team. Lyons knew that a well-engineered and designed new engine would do the job, but to fill the order book the new engine should not only perform in a spectacular fashion but should also look glamorous and exotic.

In 1946, the design team of Lyons, Heynes, Claude Baily, Harry Munday and Wally Hassan started work in earnest. The engine needed to be powerful – around 160bhp was the target figure, being where the previous unit left off – and refinement was a key goal. The engines were all prefixed with an X for experimental, followed by a letter denoting which stage of development they had reached. The first engine design was a 1360cc, 4-cylinder unit, coded XA, but all variants up to the XE were stillborn and never left the paper design stage. The XF was still rated at 1360cc (66.5mm bore x 98mm stroke) but, though not a bad engine, was never really going to have the bottom end torque required of a Jaguar unit.

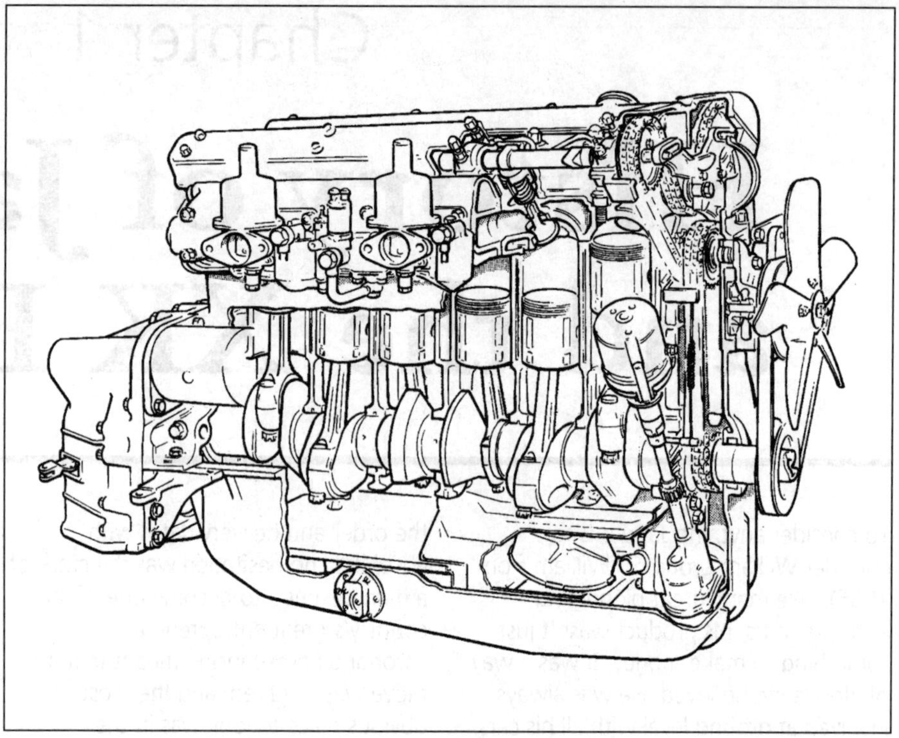

▲ *The eventual XK design, with a cast-iron block, aluminium cylinder head and a 7-bearing crankshaft, at the front of which was a gear pump which lubricated the con rods. The cams were driven by two (top and bottom) duplex timing chains. The alloy pistons were fitted with chrome top rings and the 12-valve cylinder head had an 8.0:1 compression ratio with hemispherical combustion chambers, courtesy of cylinder-head guru, Harry Weslake. These permitted excellent control of the fuel/air mixture which led to a relatively high-revving engine, producing plenty of power and torque but with excellent fuel economy (especially way back in 1948). It looked exotic and futuristic, with its distinctive polished alloy head displaying the fact that it clearly had twin overhead cams – something many engines still don't have in the 1990s; and in the late 1940s it was absolutely outrageous! You might have expected to flip the bonnet of a Grand Prix car and see those matching polished cam covers, but surely not on a road car – especially one which was massively cheaper than Italian exotica.* (Courtesy Jaguar-Daimler Heritage Trust)

The XG engine was also 4-cylinder but, being based on the existing Standard engine, had a capacity of 1776cc (73mm bore x 106mm stroke). Though this engine was put through many thousands of miles of testing (mostly by Hassan) it was still deemed to be too small and noisy and generally insufficiently refined for a Jaguar application. Part of the package was BMW-style cross pushrods, Weslake's hemispherical combustion chambers and a single overhead cam. The engine itself didn't make the grade, but all were impressed by the head and its design. So it was decided that, whatever the eventual layout, the Weslake head would be part of it.

The XJ design saw the team almost reaching their goal, as it was produced in both 4-cylinder 1996cc (80.5mm bore x 98mm stroke) and 6-cylinder 3186cc (83 mm bore x 98mm stroke) formats. Though still too lumpy and rough for serious consideration, the bigger engine in particular was much more like it, being considerably smoother and producing plenty of power. What still wasn't present in sufficient quantities was the low-down torque required of a Jaguar, and worse, it actually produced less than the outgoing 4-cylinder 3.5-litre engine.

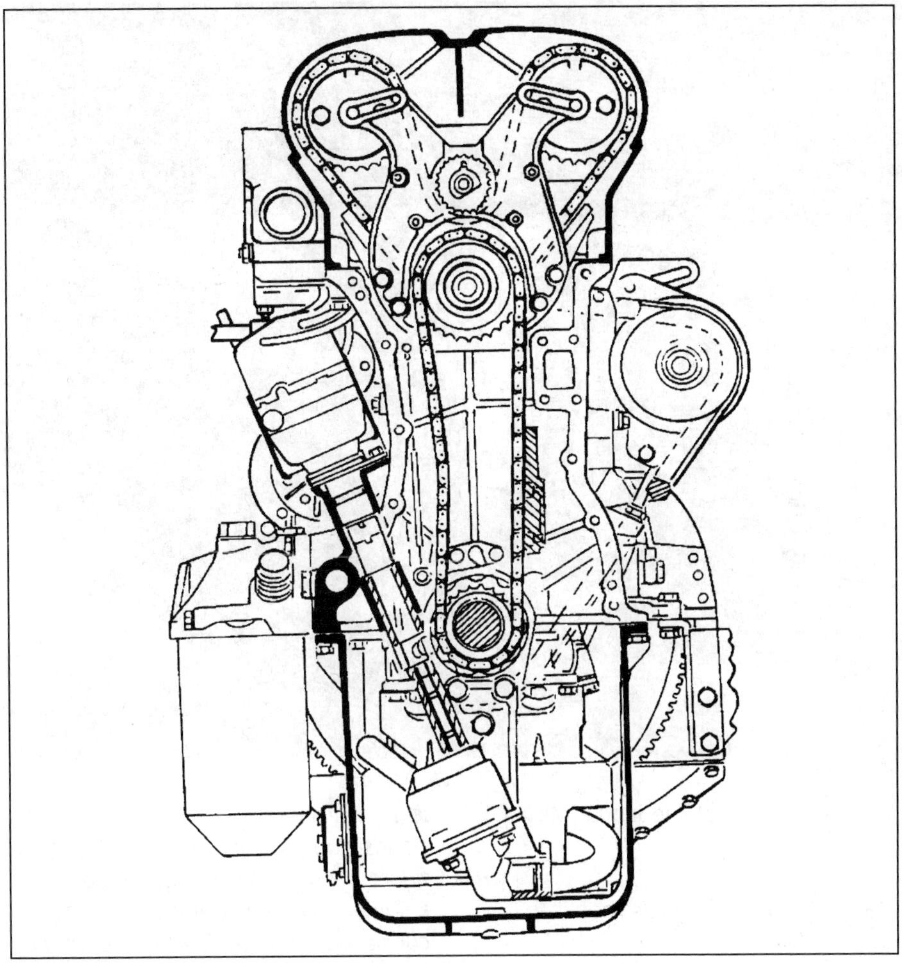

▲ *In this end-on view, the layout of the cams and chains can clearly be seen. Note the spring and blade type chain tensioner used on early cars, but soon replaced by a hydraulic version which was much more efficient and reliable. Many think that the XK engine was the first dohc cam unit to be fitted to production road cars, but they'd be wrong. The car that can lay claim to that honour is the Salmson D-type (yes, really!), produced in 1922. However, it has to be said the XK was slightly more successful.* (Courtesy Jaguar-Daimler Heritage Trust)

However, the 4-cylinder unit was certainly not without its merit, as proved by A. T. 'Goldie' Gardner, a well-known 'character' of the time, who's penchant was finding records to break. In his EX135 car (a pre-war MG chassis, streamlined for extra speed) he broke several speed records achieving a 2-way average speed of just over 176mph.

The problem of producing more torque from the 6-cylinder engine was remedied by the simple expedient of increasing the stroke to produce a capacity of 3.4 litres (a bore/stroke of 83mm x 106mm to create an actual capacity of 3442cc). The power output was rated at 160bhp at 5,000rpm and there was a healthy 195lb/ft of torque at a suitably low 2,500rpm.

At first, there was some doubt in the team that the engine would work as an everyday road car unit – a dohc unit, with (relatively) noisy twin timing chains would be more complex and costly to produce and potentially more troublesome. However, the dohc design wasn't really a surprise, as Bill Heynes was a great fan of motorcycle racing and had been greatly impressed by the twin-cam engines he saw at Brooklands, units which also featured hemispherical cylinder heads.

IMPORTANT JAGUAR DATES AT A GLANCE

1923 William Lyons, in partnership with William Walmsley, forms The Swallow Sidecar Company.

1928 The company moves to Coventry, or more precisely to the town of Foleshill slightly to the north-east of the city.

1931 The first car to bear the SS nomenclature – the SS1 – appears at the London Motor Show.

1935 William Lyons registers the name Jaguar as a name for future cars.

1945 Because the initials SS had acquired unpleasant connotations during the 1939–45 war, the company name is changed to Jaguar Cars Limited.

1948 The XK 6-cylinder engine is seen for the first time in the XK 120 sports car, launched at the London Motor Show.

1949 XK 120 makes history when it exceeds 120mph on a (closed) motorway at Jabbeke, Belgium.

1950 The Jaguar Mk VII is launched.

1951 The C-type Jaguar wins at Le Mans, the first of many such victories for Jaguar. The company moves to its present home in Browns Lane, Coventry.

1953 Jaguar triumphs again at Le Mans, as well as gaining second and fourth places as a bonus. The success owes much to their use of disc brakes.

1954 The Jaguar Mk VIIM and the XK 140 sports car are launched.

1955 The D-type wins at Le Mans in a tragic race. Jaguar pulls out of Le Mans competition until 1960. Jaguar's first compact saloon is launched, known retrospectively as the Mk 1.

1956 William Lyons becomes Sir William in the New Year Honours. An Ecurie Ecosse D-type wins at Le Mans. The Jaguar Mk VIII is launched.

1957 The Ecurie Ecosse team once more takes a D-type to a Le Mans win. The Jaguar XK 150 is launched and production of the XK SS begins, making road cars from racing D-types. A savage fire at the Browns Lane factory causes £3.5m worth of damage. The effect on one of Britain's prime post-war companies is enough to warrant a letter of condolence from the Queen.

1958 The Jaguar Mk IX is launched.

1959 The Jaguar Mk 2 is launched.

1960 Jaguar buys out the Daimler company, gaining production space and a series of V8 engines. The 2.5-litre is tried briefly in the Mk 2. A prototype E-type (the E2A) is entered for Le Mans.

1961 The E-type Jaguar is launched to world-wide acclaim at the Geneva Motor Show. This is the last Jaguar sports car to use the XK engine. The Jaguar Mk X saloon is launched.

1963 The Jaguar S-type saloon is launched.

1966 The British car industry is in disarray as Jaguar becomes part of the British Motor Corporation, the merger forming British Motor Holdings.

1968 The industry's troubles continue as BMH merges with the Leyland Group to become British Leyland. The impressive XJ6 saloon is launched to universal acclaim. Waiting lists form around the world and it becomes Jaguar's best-selling saloon.

1972 Sir William Lyons retires.

1975 The ultimate ignominy for Jaguar, as British Leyland is nationalised.

1984 Jaguar Cars is privatised again and the company begins to move once more into its ascendancy.

1985 Sir William Lyons dies at home on 8 February.

▲ *Sir William Lyons (knighted in 1956) stands alongside another famous Jaguar figure, John Egan (also later to be knighted). On the right is the stylishly curvaceous 1937 SS Jaguar, and on the left is the then current XJ6 model, clearly an S3 by the American-inspired plastic bumpers. The XJ6 was the last of the Jaguars fitted with the XK engine, which took the production run to almost 40 years.*

1988 After a gap of more than 30 years (1957) a Jaguar once more triumphs at Le Mans – this time with the XJR-9.

1989 Jaguar again changes ownership, becoming a wholly-owned subsidiary of Ford GB. The company's new owners are able to supply the cash needed for the development of new models.

1990 Another Jaguar win at Le Mans.

1992 The Le Mans-based XJ 220 road car enters production.

1995 Memories of the E-type launch are evoked by the launch of the XK8 V8-powered sports car at Geneva.

Chapter 2

XK Milestones – Saloon Cars

MK VII and MK VIIM
Production total: 30,997 from 1950 to 1957

The Mk VII was shown at the 1950 Earls Court Motor Show and the first new Jaguar saloon since the war was an instant hit. Despite its weight and bulk compared to the XK 120 (over 1.5 tons and measuring 16ft 4in) the powerful (160bhp) and torquey XK engine could still turn in a respectable 100mph top speed and a 0–60 time of 12.5 seconds – with up to six people on board. The car was based on the earlier Mk V, with the most obvious change being the installation of the XK engine to replace the ageing pushrod unit.

Why jump from Mk V to Mk VII? It wasn't because Jaguar couldn't count, but that (much to Lyons' annoyance) Bentley had already used the Mk VI title. However, once Jaguar had created the Mk VII the boot was on the other foot as it meant that Bentley couldn't follow on a natural progression from their Mk VI.

Much was made – at home and abroad – of the XK engine's lazy nature (no-one wants a high-revving buzz-saw engine in a large luxury car) and the company pointed out that at 70mph, the tachometer would be indicating a mere 2,500rpm. The combination of style, luxury (as ever, leather and wood were much in evidence inside the car) and that superb engine led to the car being a huge sales success, especially in

▲ *The 3½ dropnead was a lovely looking car, but at 160bhp …*

▼ *… the 4-cylinder engine had reached its peak. Note the SU carburettors bolted directly to the cylinder head without a manifold. Produced between 1946 and 1949, the stylish lines were subtly altered to produce …*

▲ *... the Mk V, a car which used the engines from the 2½ and 3½ models in a last gasp for the outdated unit. It gave way to one of the most successful engines of all time, one which went on to power almost all Jaguar cars for nearly 40 years.* (Courtesy Jaguar-Daimler Heritage Trust)

the USA where the biggest market lay. Records show that in the first year, orders worth more than $30m had been taken!

In 1953 an automatic gearbox option was offered for the American market, it being apposite that Jaguar used an American Borg-Warner 3-speed box (it was prohibitively expensive to develop its own unit). From 1954 overdrive became an option on manual gearbox cars and it was an instant hit in Europe, where petrol prices were much higher than in the States. In September 1954 the performance got even better with the advent of the Mk VIIM which had an extra 30bhp to put through its close-ratio manual gearbox. XK developments in racing and the XK sports cars were generally carried over to this car, including the fitting of the B-type cylinder head as it became available.

▶ *The 3.4-litre XK engine was designed for the large and luxurious Mk VII which was produced, along with its Mk VIIM stablemate, from 1951 to 1957. Incredibly, the staid and establishment Mk VII was the only Jaguar to win the Monte Carlo rally – in 1956! The Mk VIII appeared that year with more power from the engine (210bhp) and more luxurious fittings.* (Courtesy Jaguar-Daimler Heritage Trust)

▲ *The differences between the Mk VIII (seen here) and the Mk VIIM were mainly cosmetic, such as the two-tone paint schemes and, inside, more opulence and luxury for its cosseted owners and passengers.* (Courtesy Jaguar-Daimler Heritage Trust)

MK VIII
Production total: 6,247 from 1956 to 1959

Introduced in September 1956, the Mk VIII was very much a stopgap car, being based around its predecessor and using the same basic shell. It retained the 3.4-litre engine, though like the Mk VIIM it produced 210bhp, rather than the 190bhp of the standard Mk VII. The changes could mostly be filed under luxury appointments, with acres of chrome and seemingly whole forests used for the interior fittings.

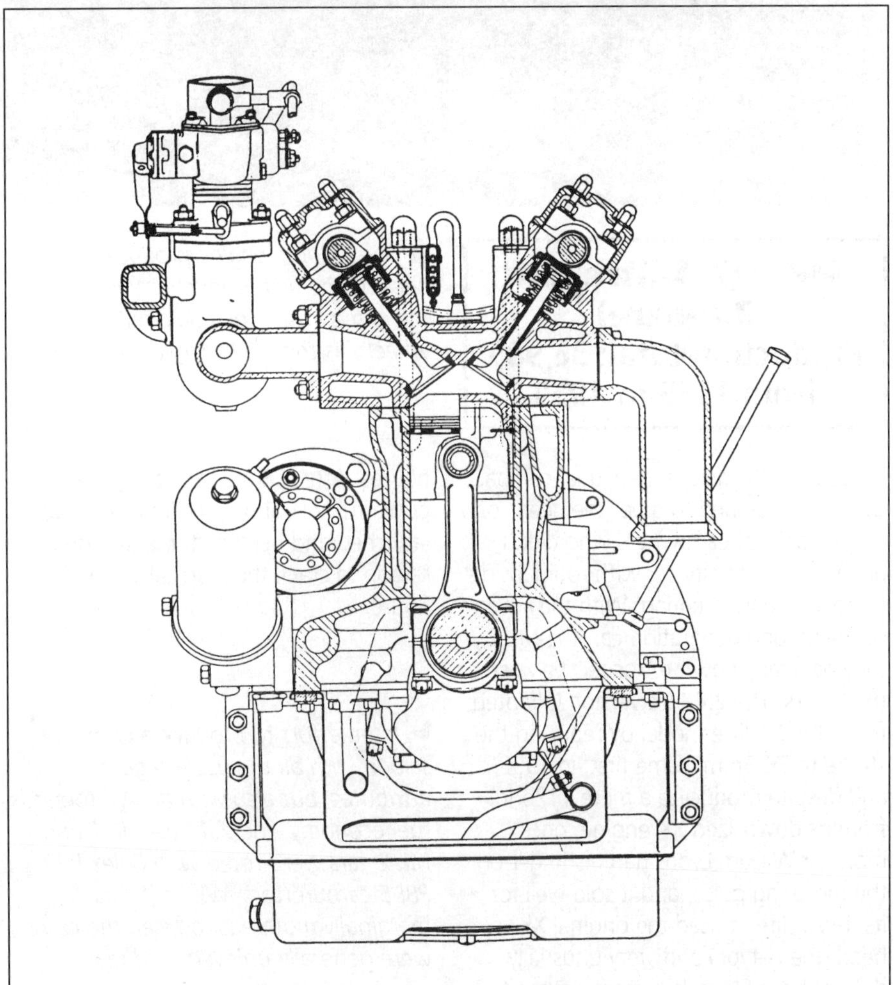

▶ *1956 was the year that saw Jaguar launching its first compact saloon, latterly known as the Mk 1. Initially, there was a choice of two engines. The 112bhp 2.4-litre is seen here in diagramatic format. Note the Solex carburettors.*

▲ *The 3.4-litre engine produced 210bhp, and with this engine installed the Mk 1 became the first production Jaguar to top 120mph. If the number plate didn't give the game away, you could tell the smaller-engined car by its fully enclosed rear wheels, as the 3.8 cars had cutaways.* (Courtesy Jaguar-Daimler Heritage Trust)

MK 1 (2.4-litre and 3.4-litre)
Production total: 36,985 from 1955 to 1959

These were the first of the Jaguar compact saloons developed to meet the needs of the majority of car owners who didn't need a large, thirsty car with room inside to hold a board meeting. With the advent of the second generation car (the Mk 2), they became known retrospectively as the Mk 1s. The 2.4-litre version (adapted from the 3.4-litre model by reducing the stroke to 76.5mm) came first, in 1955, and despite producing a mere 112bhp from its downsized XK engine, once more Sir William Lyons had his finger on the motoring pulse, and it sold well for its 4-year life. It used the original XK head (the A-type) and, very unusually, did not have SU carburettors, rather it used Solex in the pursuit of better mpg figures. After 18 months, the more powerful (210bhp) 3.4-litre model was launched and became the first Jaguar saloon to crack the magical 120mph figure.

▶ *In an effort to produce a compact saloon with all the usual Jaguar attributes, but also with an eye towards fuel economy, the 2.4-litre Mk 1 and Mk 2 cars were fitted with Solex B32 PBI-5 carburettors. Whilst it was marginally more economical, the carbs were generally unloved, and many owners simply do the swap to suitably sized SU carbs.*

MK IX
Production total: 10,002 from 1958 to 1961

But Jaguar didn't abandon the large saloon market, and though the Mk IX was clearly a derivation of the Mk VIII it broke new ground in several directions. It was the first Jaguar to feature the larger capacity 3781cc XK unit rated at a nominal 3.8 litres (the 3.4-litre Mk VIII was actually produced until the end of 1959, albeit not in any great numbers). Power and torque were up over the smaller engines, being 220bhp and 240lb/ft respectively. It could race to 60mph from rest in a smidgen over 11 seconds and, more important, could scrub off the speed with equal aplomb, being the world's first luxury saloon car to be fitted with disc brakes all round.

▲ *Though almost the same as its predecessor in many respects, the Mk IX scored two firsts; the first Jaguar with standard power steering and the first production car to be fitted with disc brakes as standard – a legacy of Jaguar's Le Mans racing programme.* (Courtesy Jaguar-Daimler Heritage Trust)

MK 2 (2.4, 3.4, 3.8-litre)/240 and 340
Production total: 83,201 from 1959 to 1967

The crisp and stylish lines of the Mk 2 and the presence of the XK engine up front – a choice of three, no less – meant instant success for one of the most popular Jaguar saloons ever. In fact, they were second only to the XJ6 which followed on from their demise in 1968. Clearly developed from the Mk 1 compact saloons, the company addressed the criticisms and equipped the Mk 2 with a range of three XK engines. The 2.4-litre remained, but with a power output increased to 120bhp, thanks to the fitting of the B-type cylinder head. The output of the 3.4-litre engine was upped to 210bhp and the 3.8-litre engine used in the Mk X easily made the transition to the smaller car. With 220bhp on tap, the performance was understandably impressive, with 60mph coming up from rest in 8.5 seconds and the top speed being a healthy 126mph. Unfortunately for Mk 2 enthusiasts, the 265bhp S engine from the XK 150S wouldn't fit because of the carburettors fouling the engine bay.

▲ *In 1959 the Mk 2 arrived, and over its production life (1959–1967) was available with 2.4, 3.4 and 3.8-litre engines, this car being powered by the latter. All Mk 2 models (not just the hot ones) featured cutaway rear wheel spats, regardless of whether wire or steel disc wheels were fitted.*

Nevertheless, the prodigious power and torque available from the twin-carburettor car made it necessary for a limited-slip differential to be a standard fitment, to give drivers of less than racing standard a chance of keeping it in a straight line!

With the advent of the 4.2-litre engine, the 3.8-litre Mk 2 models were deleted in 1967 as the range was updated and they were renamed 240 and 340, with 2.4-litre and 3.4-litre engines respectively. (Officially, there was no 3.8-litre option, although nine cars were built to special order.) The specification level was deliberately kept relatively low so as not to invite comparison with the upmarket big luxury saloons in the range at the time (the 420 and the S-type).

▲ *Seen out of the car the engine is even more impressive, and with this 220bhp lump up front the Mk 2 could hit 126mph. Not immediately obvious in a black and white photo is the blue cylinder head. Impressive at the time, the Mk 2 has gone full circle from total respectability to bank-robbers' favoured getaway car, to forgotten hero to current serious classic. Did anyone say Inspector Morse?* (Courtesy Rob Beere Racing Services)

◀ *The 240 and 340 models were introduced in 1967 as mildly re-worked Mk 2 cars, and it's hard to tell them apart.* (Courtesy Jaguar-Daimler Heritage Trust)

▶ *No matter how you twist the page, you can't make this add up to 6-cylinders in-line. Having taken over Daimler, Jaguar slotted that company's 2.5-litre V8 under the bonnet of the Mk 2. It didn't last, as it was thought a little too rough and rumbly to grace a Jaguar.*

MK X/420G
Production total: 18,633 from 1961 to 1966

Still intent on producing a top-range, large luxury saloon, Jaguar couldn't have made the Mk X much bigger – at 16ft 10in long and a staggering 6ft 4in wide it was the widest car on UK roads – until the introduction of the Jaguar XJ220! However, it wasn't aimed specifically at UK roads, rather at those multi-lane highways in the USA, where such dimensions are commonplace. The 3.8-litre engine, complete with three SU HD8 carburettors (as used in the E-type), was fairly lost in the massive engine bay, although from the start it was mooted that there should be a V8 or even a V12 power plant available. In 1964 the 3.8-litre unit was replaced by the 4.2-litre engine, which produced the same power (265bhp) but more torque, at 283lb/ft. It shares the honour (with the E-type) of being the first car to benefit from the largest ever XK engine. In 1967 the 420G model was launched, the G standing for grand and the main differences being cosmetic and luxury. It should not be confused with the 420 model, which was different altogether.

▲ *Also in production at the same time as the Mk 2 (and later, the S-type) was the none-too-slender Mk X. And the caption? 'I think we should sell the house, darling, and live in the car – there's more room!'* (Courtesy Jaguar-Daimler Heritage Trust)

▲ *Gilding the lily, the 420G added yet more luxury where most would have not thought it possible. The name was confusing, as it was not related (directly, at least) to the 420 model. A leviathan of a car, the Mk X/420G held the somewhat dubious record of being the widest UK production car until the introduction of another Jaguar – the XJ220!* (Courtesy Jaguar-Daimler Heritage Trust)

▶ *The Mk 2 saloon's successor, the S-type was the final metamorphosis of the line before the introduction of the incredibly successful XJ6.* (Courtesy Jaguar-Daimler Heritage Trust)

S-TYPE/420
Production total: 24,993 from 1963 to 1968

The compact saloon market was a healthy one for Jaguar, but as the company's luxury reputation continued to gain pace, a demand grew for a more luxurious Mk 2. Hence we come to the S-type, based on the Mk 2 but with a redesigned and larger rear end, and the installation of Jaguar's independent rear suspension, already used to good effect on the E-type. The car was available with 3.4 and 3.8-litre engines.

The Jaguar 420 is often confused with the 420G, but the two couldn't be more different. The 420 was a more luxurious S-type and, along with a host of cosmetic improvements, was fitted with the twin-carburettor 4.2-litre XK engine. Note that Mk 2 production overlapped S-type production for some while.

▲ *This is a 1968, 3.4-litre car and shows just how well the influences of the Mk 2 and the 420 have come together. The wire wheels add a touch of feline ferocity to an undoubtedly beautiful shape.*

◀ *The same, but different. With a squared-off front and redesigned rear quarters (with independent rear suspension) the XJ6 drew ever closer in the 420.* (Courtesy Jaguar-Daimler Heritage Trust)

XJ6 SERIES 1
Production total: 79,273 from 1968 to 1973

In saloon car terms, the XJ6 represented the zenith for Jaguar, taking the company into the luxury car stratosphere where, on more than one occasion, the car was compared (favourably) with Rolls-Royce. Moreover, it rationalised Jaguar's wide range of saloons into just one model. The XK engine was not an immediate choice for the car. Several engines were mooted. A V8 was tried (because of the Daimler connection where there was already plenty of experience and expertise in V8 production available) but rejected because it was not smooth enough. There was a 3-litre, 6-cylinder engine on the stocks but, although it produced almost as much power as the 4.2-litre engine, it lacked real torque, a key factor in any Jaguar saloon.

Hence, two engines were fitted, both XK – the established 4.2-litre unit (necessitating a bonnet bulge) with a choice of compression ratio (7, 8, and 9:1) with power outputs ranging between 245bhp and 225bhp. There

▲ *The press and public raved about it, and in the decade that was backing Britain it seemed most appropriate that Jaguar should unveil its finest offering to date. A glance back at the Mk X shows many similarities – four headlamps in line, large central rectangular grille, low chrome bumper and lower wing-mounted combined side/indicator lamp clusters. However, the proportions of the new car were far more pleasing and realistic. A reduction in the number of curves also gave the XJ6 a more modern appearance, an important factor in 1968. The range was initially available with both 4.2-litre and 2.8-litre engines, but the latter was never popular and always problematical, and following low sales it bit the dust in the early '70s.* (Courtesy Jaguar-Daimler Heritage Trust)

▲ *Instantly recognisable as an XK engine, it's also quite clearly from an XJ6 – those black and white ribbed cam covers are famous the world over. This is an early 4.2-unit built to UK specification, complete with SU carburettors.* (Courtesy Jaguar-Daimler Heritage Trust)

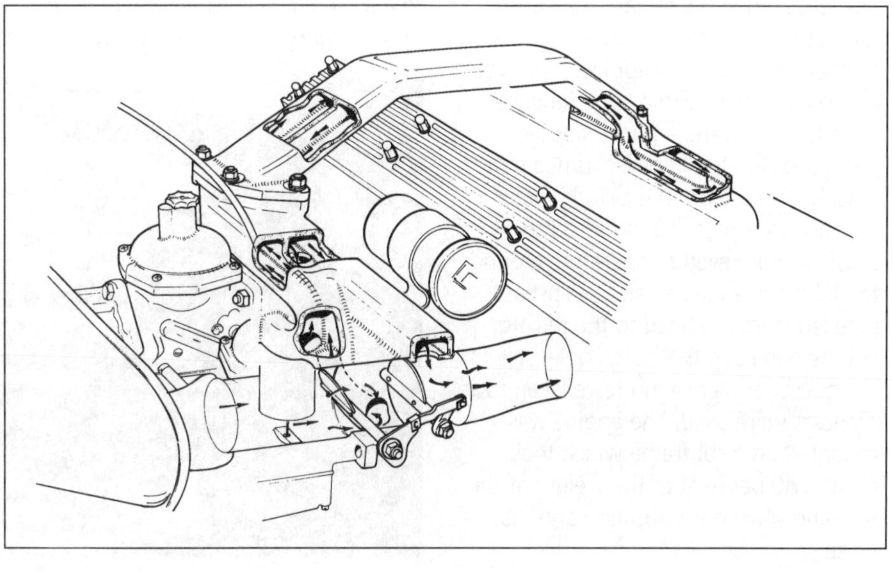

▶ *In America, Jaguar buyers were not so fortunate, as the emissions regulations dictated that twin Stromberg carburettors should be fitted and a system devised whereby the engine should eat at least some of its own emissions.* (Courtesy Jaguar-Daimler Heritage Trust)

were no qualms about performance, with a 0–60mph time of 9.7 seconds and 100mph coming up in 26.4 seconds.

The USA only ever got the 4.2-litre cars, albeit fitted with twin 175CD Stromberg carburettors, replacing the European SUs, in order to lean the mixture and reduce emissions. However, this also resulted in generally hotter running and a reduction in power, which was typically 170bhp.

Also on offer was a new engine capacity of 2.8 litres, introduced solely to beat European tax laws penalising large-engined vehicles. Developed from the existing 2.4-litre engine, the 2.8-litre XK was unloved from the start and the bigger-engined car outsold it 3:1. With the same compression ratio options, the power output varied between 160–180bhp, which really wasn't enough for such a large car.

Its troubles compounded when it developed a reputation for holing pistons, something eventually rectified (from 1970) by fitting heavier pistons – the downside being that it made the car even slower. Kept on in Europe until the Series 2 model, it was deleted from UK listings in 1972.

Following overheating problems with the Mk X and 420/420G, Jaguar uprated the cooling system. A 12-bladed viscous fan was used which operated constantly up to around 2,500rpm, whereupon it effectively freewheeled and left the engine to be cooled by the rush of air. The large crossflow radiator had a separate header tank in order that the bonnet line could be kept low. The water pump was fitted with a larger impeller – to shift more water – and a smaller pulley, so that the impeller turned more quickly. In addition, the cylinder head porting was uprated in an effort to create a more even temperature around the inlet and exhaust valves. A bypass pipe led from the head to the radiator and helped keep the head cooler still. In the quest for silence, no fewer than four silencers were used. The engine was mounted on a subframe which took around 80 per cent of the weight of the unit and soaked up vibration and road noise.

XJ6 SERIES 2
Production total: 83,318 from 1973 to 1979

USA emissions regulations meant that the 4.2-litre engines had to be fitted with a different air cleaner and an air-intake system which used exhaust heat to reduce pollution. On the LWB cars, the engine gave 170bhp at 4,500rpm and 231lb/ft of torque at 3,500rpm (the power output figures at this time changed to DIN readings, hence the apparent loss of power compared with the earlier models). In 1978, American cars were fitted with Lucas-Bosch

▲ *The Series 2 XJ6 cars benefited from a longer wheelbase (the same as the Series 1 LWB versions) and a myriad of modifications, not least of which was a rise in bumper height to come into line with USA safety regulations.* (Courtesy Jaguar-Daimler Heritage Trust)

▼ *Jaguar continued modifying and uprating the XK engine throughout the life of the XJ6. This is a concours example of a Series 3 4.2-litre unit. Right from the start there was plenty of extra room in the engine bay as it was envisaged from the design stage that it would have to take the V12 engine. Note the heavily matt-black painted exhaust manifolds, more usually coloured rust brown!*

D-Jetronic fuel injection (already tried on Jaguar's V12 engine) and 3-way catalyst exhaust systems in order to pass the ever-more stringent emissions regulations. The more efficient system enabled the designers to increase the compression ratio to 8:1 (it had been reduced to 7.4:1 during previous attempts to cut down the emissions) and thus improve the performance and power output from 161bhp to 176bhp. By the end of 1978 some European 4.2-litre cars were being fitted with the same system, albeit without the catalytic exhaust.

In April 1975 the 3.4-litre model was introduced, Jaguar looking for sales in the lucrative executive fleet market, where a smaller-engined car would be more attractive. It had the same bore and stroke as previous 3.4-litre models, but with better cooling, better internal oil flow and a totally new block – essential, as Jaguar had ditched the tooling for the original 3.4-litre unit! Two SU HS8 carburettors were linked to an automatic choke and used the straight port head from the 4.2-litre model. With a bore x stroke the same as the 2.8-litre cars (83mm x 106mm), and with a compression ratio of 8:1, it produced 161bhp (DIN) at 5,000rpm and 189lb/ft of torque at 3,500rpm.

▲ *The Series 3 XJ6 showed yet more American safety influences, with the introduction of the plastic bumpers. However, it is still distinctly a Jaguar, a symbol of Britishness so beloved across the pond.* (Courtesy Jaguar-Daimler Heritage Trust)

▲ *This is a 1983 Series 3 4.2-litre engine, by this time with a redesigned cylinder block (1979) and equipped with Lucas-Bosch L-Jetronic fuel injection (1981) to produce more power, better economy and lower emissions (North American cars had been fuel-injected since 1978).*

XJ6 SERIES 3
Production total: 123,696 from 1979 to 1987

After the quality control troubles of the 1970s, Jaguar got its commercial act together for the 1980s with the result that the Series 3 XJ6 sold 50 per cent more than its predecessor. All 4.2-litre cars were fitted with Lucas-Bosch L-Jetronic fuel injection. The compression ratio increased to 8.7:1 (from the previous 7.8:1 and achieved by use of 9:1 pistons as fitted in the E-type) which combined with earlier opening and larger inlet valves (1⅞), Lucas-Bosch fuel injection and OPUS electronic ignition to produce a power output of 205bhp at 5,000rpm. Torque was 236lb/ft at 3,750rpm but the most important aspect here was that almost all of that (231lb/ft) was available as low as 1,500rpm – only just over idling speed! The 3.4-litre cars continued mostly unchanged and had to make do with twin SU carbs.

▶ *Though inherently more complex than carburation, the fuel-injection system remained on song for longer and was considerably more efficient. Many owners feared troubles in the black box – the ECU (electronic control unit) that deciphered the information received from sensors around the engine and gave instructions as to how much fuel to allow through the injectors. In truth, most fuel injection problems relate to sensor failure than problems with the ECU itself.*

Chapter 3

XK Milestones – Sports Cars

XK 120
Production total: 12,061 from 1949 to 1954

The XK engine was originally designed and intended for use in the Mk VII saloon, but the company needed a way of publicising the engine prior to the car's launch. Publicity master William Lyons knew that the quickest way to column inches (and, ultimately, sales) was to use a sports car to grab the public's attention. Moreover, if such a car could be successfully used on the race track he knew that its reputation – and that of the engine – would rise still further. And, on a practical note, it made a great deal of sense to try out the engine in what he assumed would be a low-volume production car, rather than making mistakes on a grander scale.

Unfortunately, Jaguar didn't have a sports car available at the time. The only answer was to build one, and what an answer it was, for the XK 120 was (and is), one of the most beautiful sports cars ever made. Incredibly, it was conceived only a few weeks before the 1948 Earls Court Motor Show and, for speed and ease, was based on a shortened Mk V chassis (something that enabled much of the saloon's running gear to be used). The William Lyons design (based loosely on a 1940 BMW sports car) remains an all-time classic, a masterpiece of total rightness straight out of the box, and proof positive that he could truly be regarded as a design genius.

The effect of its launch, in open-top form, on a British public still suffering a life of bland austerity and rationing, cannot be overestimated. Even more important was its price – just £988 (although with purchase tax, this rose to a tad under £1,300 – still bargain basement, whichever way you look at it). This show car (registration number HKV 455) went on to become a mobile test bed for the range prior to the gearing up for production proper.

As a purely promotional vehicle, its production life was expected to be strictly limited (after all, the Mk VII was the important vehicle). Steel being at a premium just after the war, the first 240 cars had aluminium bodywork wrapped around a wooden (ash) frame, but this

◀ *The XK engine entered the motoring world via the stunning XK 120 sports car. Originally designed (in a matter of weeks!) as a display platform for the engine at the 1948 London Motor Show, it was understandably a world-wide success. The rear wheel spats had to be removed when wire wheels were chosen because the spinners fouled the bodywork.* (Courtesy Jaguar-Daimler Heritage Trust)

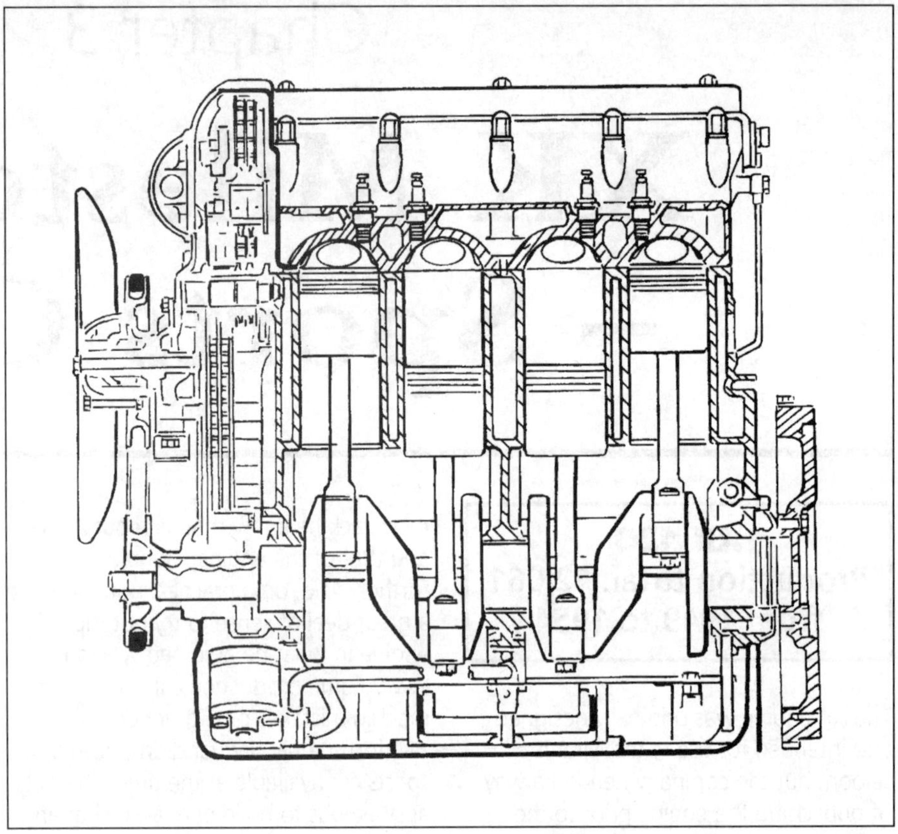

▶ *Though it seems curious now, the company believed that the large 6-cylinder engine would simply be a flagship and that the greatest demand would be for a smaller, simpler and more frugal unit. Because of this, Jaguar initially included an option in the brochures – the XK 100, powered by a 2-litre, 95bhp, 4-cylinder engine. However, the way that the XK 120 was snapped up ensured that this project was stillborn.* (Courtesy Jaguar-Daimler Heritage Trust)

soon changed, as the car-buying public took the XK 120 very seriously, and Jaguar was soon beset by long queues of eager, cash-laden enthusiasts. As a result, the XK 120 went into production proper and ultimately over 12,000 were made. However, production did not really get under way in any real sense until 1950, and even then few cars stayed at home; in the post-war export or die climate, the XK 120 was manna from heaven, with a large proportion of the cars going overseas, notably to America and Australia.

Initially, there was some understandable scepticism about the performance claims; 0–60 in 10 seconds is pretty sharp even today, and equally 120mph is not to be sniffed at; in 1948 it was the stuff of racing cars. For example, the similarly-priced contemporary Allard V8 could not crack 90mph, despite having a larger capacity engine as well as two extra cylinders. In 1947 a Healey sports saloon was taken to Belgium by its makers to convince the world (and more notably, the motoring press) of its performance capabilities. It achieved a top speed of almost 111mph which was mightily impressive. William Lyons took the same route to prove his point (to sceptical press and public alike); an XK 120 was timed by the Belgian RAC in a special demonstration run at the Jabbeke motorway – closed for the event! Driven by test driver Roland (R.M.V.) 'Soapy' Sutton, a standard road car (The debate over exactly how standard the car was still rages, with plenty of speculation that it was slightly warmed over to ensure that it hit its target. Nevertheless, its performance was still mightily impressive and cars off the production line were known to get well over the 110mph mark without too much trouble.) tripped the counters at 126.4mph, and with the windscreen replaced by a cowl and a metal cover over the passenger area (both items were listed as optional extras) the car achieved a mean 2-way speed of 132.5mph.

These figures, regardless of anything else, meant that Jaguar could lay claim to be producing the fastest production car in the world. No wonder sales took off!

In 1951 a fixed head coupé (FHC) version of the XK 120 was introduced. The SE (special equipment) model featured a hotter XK unit with 8:1 compression ratio, twin straight-through exhaust system (after six months, this was deleted and the cars were equipped with the original single-pipe systems), different crankshaft damper and ¾ lift camshafts, the main effect of the latter being to raise the power output to 180bhp or more. A lightened flywheel meant that the engine would spin more readily, and when all these factors were weighed up, quite naturally there was an increase in performance – 60mph was reached in 9.9 seconds despite the fact that the FHC was 0.5cwt heavier than the previous (all-steel) version and weighed 1.5cwt more than the original, alloy-bodied car – much of this quickness off the line can be attributed to the lighter flywheel. Only 195 FHC cars were produced in RHD form.

XK 120s were raced, rallied and hill-climbed all over the world by such notables as Ian and Pat Appleyard, Leslie Johnson and, of course, Stirling Moss. The latter used an XK 120 FHC for some time – to tow his caravan to race meetings!

In 1953 a FHC XK 120 took part in a high-speed test of endurance to prove still further (as if it were needed) that the Jaguar engine was a staggering force to be reckoned with. The event took place at the Montlhéry circuit in France where the car was driven by four racing drivers (Moss, Hadley, Fairman and Johnson) for a whole week non-stop – that's day and night, equating to 168 hours. The car covered almost 17,000 miles and, more important, its *average* speed came out at more than 100mph! More amazing still was that the test car was then driven back to the UK by the editor of a motoring magazine, who reckoned that there was no difference in performance between this and a standard XK120.

XK 140
Production total: 8,935 from 1954 to 1957

Clearly a development of the gorgeous flowing lines of the XK 120, the XK 140 was launched at the Earls Court Motor Show in 1954. Many purists rate it below the original XK sports car but, nevertheless, it managed to sell almost 9,000 units in its short (just over 2 years) production life. Even more interesting, USA buyers snapped it up more quickly than its predecessor or its successor.

Visually, a brief glance made it hard to tell the two cars apart, but most of the panels were different, as were lots of interior and exterior trim details. The main running gear change was the adoption of rack and pinion steering.

The 3.4-litre unit was effectively the SE engine from the XK 120, though extra work on the cylinder head inlet ports and attention to the ignition advance curve saw this rise by a further 10bhp over its predecessor to 190bhp. As usual, there was a performance version of the XK 140, and again, the SE nomenclature was used. This fished out another 20bhp by means of a 2-pipe exhaust system, and C-type head with bigger inlet valves and ports. Performance claims (either car) varied somewhat, and lucky buyers could expect to achieve a top speed of between 122 and 130mph. Compared to the XK 120, the engine was further forward by around 75mm, a move which enabled the DHC and FHC to become 2+2 cars, rather than 2-seaters. It also enabled a Laycock overdrive option to be offered, something not possible in the more cramped XK 120. Naturally enough, the Jaguar engineers took the opportunity of a higher top gear, to reduce the final drive ratio and so increase the acceleration of the car. Apart from this and the more relaxed high-speed cruising ability, the XK 140's longer legs meant that fuel consumption improved quite healthily on longer journeys.

When Jaguar expected a short production run for the XK 120, it couldn't have been more wrong, especially in America, where it was a huge success – so much so that after five years the company begat the XK 140; a move which was distinctly evolution rather than revolution. (Courtesy Jaguar-Daimler Heritage Trust)

XK 150
Production total: 9,712 from 1957 to 1961

The XK 150 was the final XK sports car development, and to some extent could be seen as the car growing up, looking almost saloon-like from some angles (and, indeed, weighing about the same as the contemporary 3.4-litre saloon). In a way, its development can be mirrored by the E-type in the early 1970s, as it had become heavier, less wieldy and, though more powerful, not overall a better performer.

As ever, it was available with or without a roof and initially with the engine tuned to produce 190bhp. 210bhp was available when it was fitted with the B-type cylinder head (which used C-type inlet valves and XK 140 inlet ports to bring the peak torque down by almost 1,000rpm).

In February 1959 the 3.4S was introduced. Thanks largely to an increase in compression ratio, a Weslake-developed straight port cylinder head and triple SU carburettors, this was able to produce a heady 250bhp. In S guise, the XK 150 offered truly startling performance, equalling (and usually bettering) many real exotics, and coming very close to that of the racing C-type.

In October of that year, though, the 3.8-litre engine, which had been developed through racing the D-type, was made available for the car. Unlike the racing cars, the extra capacity was not achieved by boring out the standard 3.4-litre block (which had not proved totally reliable), but by using a new block with dry liners and an increased bore of

XK 150

◀ *By 1957, the XK 120 design had progressed through the XK 140 to what was almost an XK saloon – the XK 150. This was a far more grown-up car, with more creature comforts, better brakes, handling and, of course, the latest developments in the under-bonnet department.* (Courtesy Jaguar-Daimler Heritage Trust)

▲ *If you've got it, flaunt it – the boot badge of this XK 150 reminds onlookers of Jaguar's successes at Le Mans in 1951, 53, 55, 56 and 57. Produced until 1961, it continued to sell well in the face of much-improving competition, and in 3.8S guise it featured the engine which was to power the mighty E-type.*

◀ *As with previous XK sports cars, the XK 150 was available with or without a roof, as here. The resemblance between this and the Mk 1/2 is clear to see, as it became more sophisticated and grew away from the original 'two folks and a powerful engine' essence of the sports car.* (Courtesy Jaguar-Daimler Heritage Trust)

87mm. For the XK 150, it used the twin SU HD6 carburettors of the normal 3.4-litre cars and produced 220bhp. Triple SU HD8 carburettors were available for fitting to the larger engine and, together with a Weslake-designed (of course) straight-port cylinder head, were instrumental in producing a massive 265bhp for the XK 150S.

It's worth noting that the XK 150's performance could be used to greater effect, despite its extra weight, because it was equipped with Jaguar/Dunlop-developed disc brakes, a true case of racing improving the breed. In addition, all S-type XK 150 models benefited from a couple of standard extras; the 5th gear overdrive, giving quieter and more relaxed high-speed cruising and better fuel consumption, and in an effort to tame the effects of all that power charging through the rear wheels, a Thornton Powr-Lok limited slip differential was fitted.

By 1961 the XK 150 had clearly reached the end of its development life (though the same could not be said of the engine) but its successor was waiting non-too-quietly in the wings and was destined to create even more motoring shock waves than the XK 120 had done 13 years earlier.

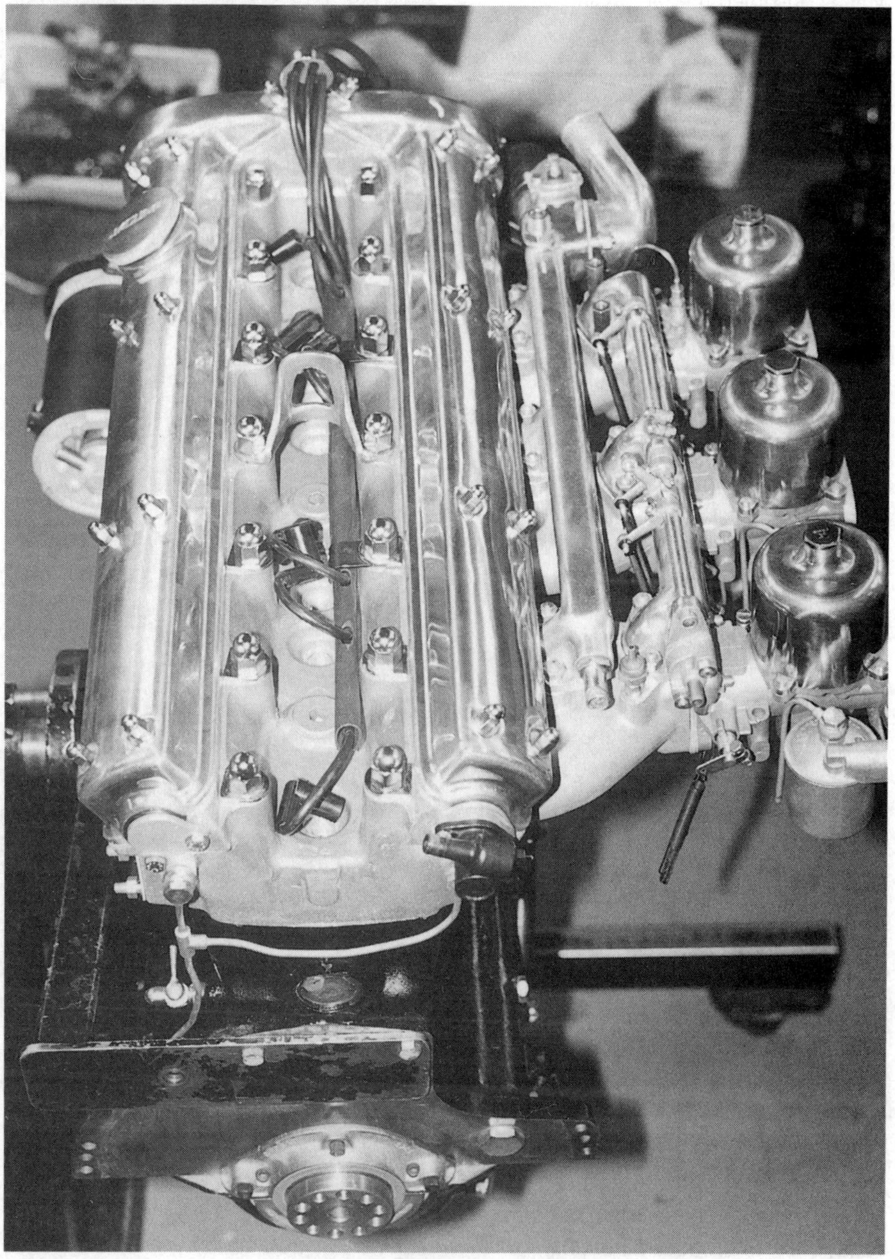

▶ *Made to make your mouth water, this is a 250bhp XK 150, 3.4S engine, as rebuilt by Rob Beere Racing Services. Note the three HD6 SU carburettors hanging on the side, when two were the norm. Though standard 3.4-litre XK 150 cylinder heads were painted duck egg green, the S versions (3.4-litre and later 3.8-litre) were painted gold, as here. (Eagle-eyed readers will spot the non-standard lip seal modification, shown in more detail later.)* (Courtesy Rob Beere Racing Services)

XK SS
Production total: 16 in 1957

Never an official production car as such, the XK SS was little more than a D-type racer civilised for road use in order to use up the stock of surplus cars/parts. However, the underpinnings remained pure D-type and the performance was, as you might imagine, absolutely sensational. The limited production run was halted for ever by the terrible fire which consumed the factory in 1957.

▲ *The XK SS had a 0–60mph time of less than 6 seconds which is seriously quick even in the '90s, let alone in the late '50s! It's hard to imagine why enthusiasts weren't queuing half way round Coventry to get their hands on one of these – certainly, they are much in demand today.* (Courtesy Jaguar-Daimler Heritage Trust)

E-TYPE SERIES 1
Production total: 38,389 from 1961 to 1968

The true road-going derivation of the D-type was, not unnaturally, the E-type. Launched at the 1961 Geneva Motor Show, the E-type is, for many, the archetypal Jaguar sports car – certainly, it has to be the most easily recognised Jaguar ever built. It used what was effectively the 3.8-litre engine from he XK 150S (but with a manual choke), and putting out the same 265 horses. All 6-cylinder E-types came with a 9.0:1 compression ratio, although an 8.0:1 version was on offer for those who couldn't get access to top quality fuel. Its top speed of 150mph was massaged slightly for the benefit of good publicity, not least by the use of lightweight panels in the road test cars, but this detracts not one jot from that stunning shape and undeniably staggering performance. Like its XK 120 ancestor, the Americans loved it, and over two thirds of production went Stateside.

In October 1964 the car was given a mild update (which included an all-synchromesh gearbox and better brakes) but the most important factor was the E-type became the joint first recipient (with the Mk X) of the largest capacity production XK engine. The block was redesigned to produce a nominal capacity of 4.2 litres. Power was pegged at 265bhp (indeed, no E-type 6-cylinder engine was quoted as giving more), but the torque was usefully improved to 283ft/lb. Early cooling problems were solved by fitting the car with two cooling

▲ *From any angle, it is a beautifully brutal sports car and looks exactly like what it is – a racer for the road. The side-exiting exhaust leaves no doubt as to its true intent.* (Courtesy Jaguar-Daimler Heritage Trust)

fans (rather than the single unit of the first models), and by increasing the amount of air flow to the radiator.

In 1966 the wheelbase was stretched by 9in to make the hard top version a 2+2 – though the space behind the rear seats was really suitable only for small children. Some said that the extra length spoiled the overall effect of the car, but most agreed that it still looked incredible from any angle. The penalty for extra interior space was, of course, extra weight – around 100kg. In the UK it has always been the least popular XK-engined E-type and is currently the most economic method of buying into the E-type market. However, in the USA the situation is reversed, as the Americans always loved the 2+2.

Technically, it was the same as the rest of the range (all now had the 4.2-litre engine), although there was an auto option available which, as ever, sapped performance so that the 0–60mph yardstick lengthened to 8.9 seconds.

▲ *The E-type roadster and fixed head coupé were introduced at the March 1961 Geneva Motor Show. The prices were a tad under £2,200 for the coupé and around £100 less for the roadster. Those were the days!* (Courtesy Jaguar-Daimler Heritage Trust)

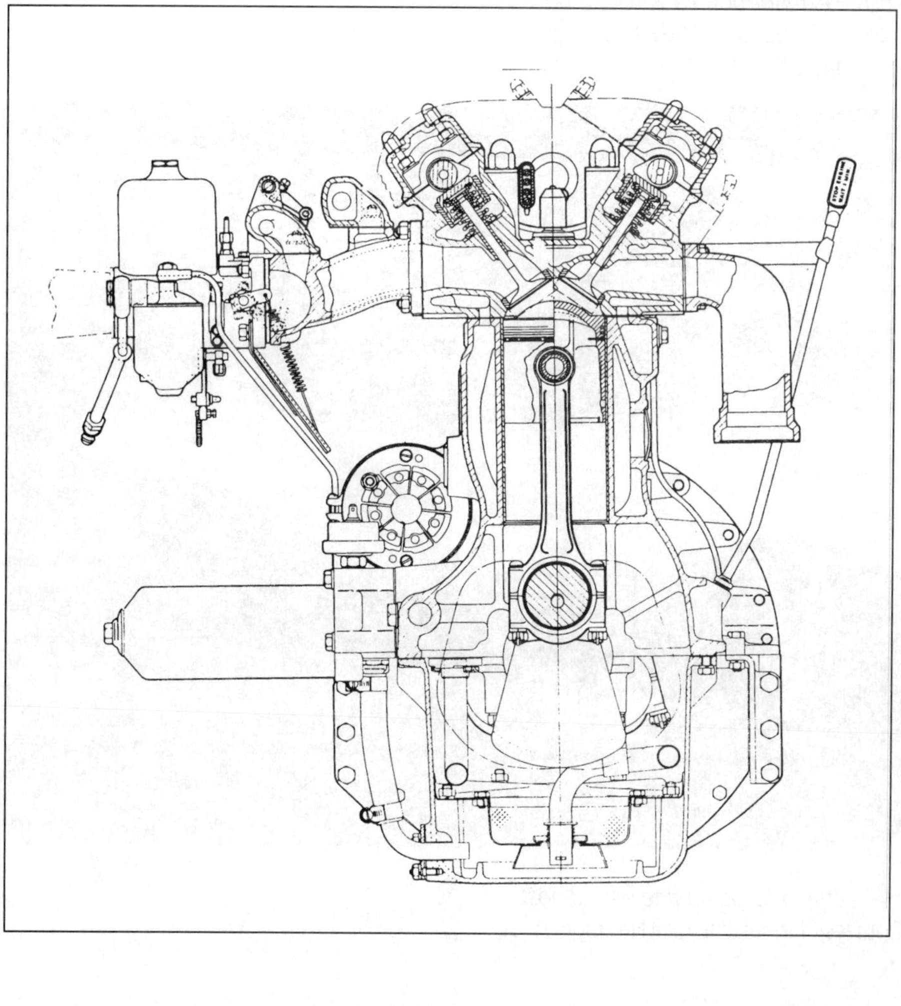

◀ *This is the E-type XK engine in cross section. Note the horizontal oil filter canister.*

▲ *The most famous 'cat' in the world? Whatever, there can't be many who don't recognise the E-type. The faired-in headlamps, small bumpers and a host of other details denote this as a Series 1 car. Moreover, the exterior bonnet catches show that this is quite an early model, as they were deleted in September 1961 to be replaced by internal catches.* (Courtesy Jaguar-Daimler Heritage Trust)

▲ *Thought by many to be the best E-type made, the original engine was almost exactly the same as the 3.8-litre XK 150S.* (Courtesy Jaguar-Daimler Heritage Trust)

▶ *The main difference was that the choke became manual for the E-type and the air cleaner assembly was slightly different. The three carburettors are SU HD8, 2in models.* (Courtesy Rob Beere Racing Services)

▶ *From any angle it's a lovely piece of machinery. It's a pity those exhaust manifolds won't stay untroubled by rust (although companies such as SC Parts now offer a stainless steel header to go with its range of stainless exhaust systems).* (Courtesy Rob Beere Racing Services)

▲ *The Series 2 cars were introduced in 1968, though the distinctive lack of faired-in headlamps actually first appeared (or rather, disappeared!) in 1967 on the so-called unofficial Series 1½ cars.*

▲ *The 4.2-litre engine was introduced for the E-type in 1964. Though UK models retained those three large SU carburettors, North American cars were stuck with just two Stromberg carburettors as a sop to the increasingly stringent emissions regulations. Naturally, this had a detrimental effect on the power output, typically cutting it by 40–50bhp. Note the (now) trademark black and white ribbed cam covers and that all engines with these covers featured unpainted cylinder heads.*

E-TYPE SERIES 2
Production total: 18,848 from 1968 to 1970

An obvious development of the earlier cars, American safety and emissions legislation was the main factor in its weight gain and design changes. Indeed, by the time the Series 2 had finished production, American cars, complete with Stromberg rather than SU carburettors, were producing just 177bhp – some 80bhp down. It was still available in roadster, fixed head and 2+2 coupé formats. The Series 2 models were only ever fitted with the 4.2-litre engine, still quoted at 265bhp at 5,400rpm and 283lb/ft of torque at 4,500rpm.

◀ *The 3.8-litre cars were fitted with poor cooling systems and only a single, low-powered fan. Jaguar took the opportunity with the 4.2-litre models to address the problem by fitting twin fans.*

▼ *The Series 3 E-type, introduced in 1971, and its successor the XJS, used the 12-cylinder, 5.3-litre engine. By 1970 the car had almost run its course and, more important, it was the last Jaguar sports car to feature the XK 6-cylinder engine. The Series 3 was undeniably an E-type but, for many, its extra weight and girth took it too far away from the original concept.* (Courtesy Jaguar-Daimler Heritage Trust)

▲ *The larger, gaping grille also came in for criticism, although many (present company included) believe that it adds a certain feline aggressiveness to the car, especially with the larger wheel and tyre combination and the flared arches they made necessary. Still, you pays your money, you takes your choice, and though not an XK, the V12 engine was (and is) regarded as one of the finest in the world.* (Courtesy Jaguar-Daimler Heritage Trust)

Chapter 4

The XK's Track Record

It's true to say that competition improves the breed. It's equally true to say that *winning* in competition improves the image and, ultimately, the sales figures.

The XK 120 was, naturally enough, the first XK-engined Jaguar to hit the race tracks. After an initial shakedown by 'Lofty' England and Walter Hassan, Jaguar (unofficially) fielded an impressive team (Prince Bira of Siam, Peter Walker and Leslie Johnson) for the first outing at the One Hour Production Car Race at Silverstone in August 1949. The history books were robbed of a fairy-tale 1–2–3 XK finish when Prince Bira's car blew a rear tyre, leaving Johnson to come home first, closely followed by Walker.

However, few would complain at a 1–2 finish first time out! It is interesting to note that the first XK 120 to leave the production line, registered as HKV 455, was the one used by Prince Bira on this occasion.

In rallying terms, the most success went to Ian Appleyard with his XK 120 registered NUB 120 – a famous number and one seen on innumerable photos. Ian had more Jaguar connections than most, for as well as being an accredited Jaguar dealer, his co-driver was his wife, and she just happened to be William Lyons' daughter, Pat!

In 1950 William Lyons employed the famous Italian driver, Clemente Biondetti to drive an XK 120 in the historic Targa Florio road race in Sicily. He had already won it twice (in 1948 and 1949) but he was denied a hat trick, as his car rolled to a smoky demise. It appears that a con rod had cracked, wrecking the engine. The result of this was that all production con rods were crack-tested.

The Mille Miglia took place less than a month later and there were four XK 120s in the hunt, although all were massively overweight compared to their Latin opposition. Biondetti came in eighth after struggling for most of the race with an overheating engine. Nick Haines crashed and Tommy Wisdom's gearbox jammed, but Leslie Johnson made fifth place. On the face of it, not an outstanding result, but the four cars ahead of them comprised three Ferraris and an Alfa Romeo with a certain Juan Manuel Fangio at the helm! As ever, all things have to be seen in perspective.

In 1950 Jaguar entered Le Mans. In essence, the entries were little more than slightly modified production XK 120s. William Lyons was well aware of the massive publicity to be gained by winning at Le Mans, but also knew he was unlikely to win first time out. Far better, he reasoned, to do some reconnaissance before mounting a full-scale attack.

Lightened by removing unnecessary items (passenger seat, etc) and with the necessary addition of a much larger fuel tank and safety items such as extra lighting, the cars were ready to race. Three two-man teams were entered; namely Leslie Johnson/Bert Hadley (did not finish), Peter Whitehead/John Marshall (15th) and Nick Haines/Peter Clark (12th). All complained of poor braking (already a common XK 120 ailment), but despite this and other minor troubles, two out of three cars made it to the finish line. Johnson's car was in the running for at least third place when the clutch plate broke while the car was out on the circuit. It's a point worth noting that, just as Biondetti's broken con-rod caused factory crack-testing to be introduced, this incident was pivotal to the introduction of solid-centre clutches on all racing Jaguars from that date.

Following the 1950 Le Mans effort the factory turned its attention once more to the annual Silverstone International Trophy race. Naturally, having cleaned-up quite emphatically the previous year, they were keen to repeat the exercise. The three cars recently returned from France were repaired for Johnson, Walker and Wisdom, and additional cars were driven by Peter Whitehead and Tony Rolt. Walker and Rolt came home first and second in their heat, but because the race was decided over two heats and because rain fell between the two, overall victory eluded the XK 120s.

Later that year, the cars made no mistake in Dundrod, Northern Ireland, during the first staging of the Tourist Trophy since 1938. Haines, Rolt, Walker and Whitehead were entered, but the wild (incredibly wild!) card was that Tommy Wisdom had entered an XK 120 to be driven by a very young and almost unknown English driver – an impressive youngster called Stirling Moss! In conditions where a rudder would have been more use than a steering wheel,

▲ *As well as a string of tarmac racing successes, the XK 120 was a fair performer when things weren't quite as smooth. This is Ian and Pat Appleyard in their famous NUB 120 winning the Alpine Rally in 1950. It's an evocative shot showing the Rhône Glacier as a backdrop. The Appleyards notched up an admirable string of wins in various events from the 1951 Tulip Rally to the RAC Rally in the same year and clean runs on the Alpine Rally in both 1951 and 1952 to give them the hat trick.* (Courtesy Jaguar-Daimler Heritage Trust)

the 20-year-old Moss came home first, and in the process set the fastest lap time on the last lap of the race. Not surprisingly, 'Lofty' England made sure that young Stirling's name was on a Jaguar contract for the following year.

The first event of 1951 was the Mille Miglia where the chief XK 120 protagonists were Stirling Moss and Leslie Johnson. It wasn't a good event, for both were out very early on, coming to grief on the oil spillage from the engine blow-up of another competitor.

The third Silverstone Daily Express International Trophy took place in May, rather than August, but its format remained the same, as did the need for a Jaguar win. Moss joined old hands Wisdom, Walker and Johnson. By now there was sufficient production of the car (for this country) to enable privateers to have a go – the most notable perhaps at this time being Duncan Hamilton. Walker and Moss tied for pole position time, but throttle problems in the heat for Walker left Moss untroubled in his win, with Hamilton coming in third in an almost standard, road-going car. XK 120s came fourth and fifth, though the event of the day was ex-motorcyclist, Charlie Dodson bringing his XK 120 into second place – despite not having raced for 12 years!

Five XK 120s cluttered up the top six places in the International Rally du Soleil in France, Wood took the Scottish Rally and the first place XK 120 in the Liege/Rome/Liege Rally was none other than Jacques Ickx, the father of Jackie Ickx, GP and, more famously, sports car driver and many times Le Mans winner.

The XK 140, achieved nothing like the competition prominence of its predecessor, due in no small part to the relative lack of Jaguar involvement in motor sport following the tragic accident during the 1955 Le Mans race. However, in 1956 the redoubtable Appleyards took an XK 140 to a second place on the RAC Rally, and an XK 140 won its class in the Mille Miglia.

C-TYPE

The success achieved by drivers with the XK 120 spurred Jaguar on to up the ante, with the XK 120 Competition type, or C type, as it has become known. Visually, it was clearly much more of a racing car, which was exactly its raison d'être. With its tubular spaceframe chassis and aluminium bodywork, its target was Le Mans, where in 1951 Walker and Whitehead won by miles – 67 to be precise. However, it could have been a truly glorious 1–2–3 had the sister cars of Moss and Biondetti not been sidelined by a sump oil pipe fracture. The XK engine was left mostly to its own devices, though larger (2in) SU carburettors were fitted to make the best of the higher lift cam and 1⅝ valves (together with suitably enlarged valve ports). The C-type was a second-time winner later in the year when Moss clinched victory in the TT.

1952 was not good for the C-type and Le Mans. The rumours that the Mercedes entry had a massive straight-line speed advantage resulted in some hasty revision of the bodywork to improve streamlining and thus gain more top speed. However, the main effect was overheating and, ultimately, retirement. The following year the bodywork was very wisely un-revised(!) back to the way it was in 1951, and the wisdom of this is proved by the results – first, second and fourth! Weight was saved on chassis and bodywork and, for the first time, SU carburation was ditched in favour of Webers – this has since become common practice where such modifications are allowed on racing engines. But it has to be said that probably the main contributory factor to the C-type's success that year was the fitting of disc brakes, allowing the Jaguars to brake later and harder than the drum-braked opposition.

▲ *Wouldn't you just love one of these? OK, maybe not in exactly this condition, but you can just glimpse the twin polished cam covers over the radiator, although there's little else giving clues to its identity, and it's not easy without its clothes.* (Courtesy Rob Beere Racing Services)

▲ *Better now? The tubular chassis marks it out as possibly being a racer, and the Jaguar logo in Rob Beere's workshop confirms its heritage. It is, in fact, a C-type which appeared in the workshops needing some attention – to say the least! However, its well worth the effort, as it is one of only 53 production C-types made.* (Courtesy Rob Beere Racing Services)

▲ *Fully clothed and restored, the C-type looks every inch a racer, even though it was very much a modified road car – indeed, many owners took great delight in making the most of their road-legal chargers. Only a few mods were required to make a 1951 Le Mans winner into a car fit and safe for 1950s UK roads, one of which was the addition of a horn – some may argue that a horn would have been most useful down the Mulsanne straight! And don't forget that the C-type was produced at a time when it was the norm for racing cars to be driven to an event, equipped with racing numbers and tyres, raced for the afternoon and then driven home again – accidents notwithstanding.* (Courtesy Jaguar-Daimler Heritage Trust)Heritage Trust)

D-TYPE

Not unnaturally the C-type's successor was the D-type, though this was an altogether more serious racing car, and the intent was the same – to win Le Mans. The stunning lines of the D-type were created by Malcolm Sayer, not surprisingly an aerodynamicist, who had joined Jaguar from the Bristol Aircraft Company. In 1954 a brief shakedown of the car at Le Mans saw it instantly hacking some five seconds off the 1953 lap record set by Ascari in a Ferrari – a good start. The car was lightweight, thanks to the use of 18g magnesium alloy for the central body section (the tub), and this material also gave it tremendous strength, something of a necessity when travelling at high speed for a full 24 hours. The front aluminium sub-frame completed the basic chassis, and from this description it's not hard to see the origins of the road-going E-type developed some years later.

Despite the early promise, a mysterious misfire and various other mishaps meant no maiden win, although finishing just 105 seconds (in 24 hours don't forget) behind the winning Ferrari was no mean feat for Rolt and Hamilton.

In 1955 many changes were made to the car, not least under the bonnet where the inlet valve size was increased yet again to 2in and the exhaust valve size also increased by 1/16in to 1 11/16in. The inlet valve inclination was retained at the original 35 degrees, but the exhaust valve angle was changed to 40 degrees to prevent overlap. This combination of angles became available on road-going options, where it was always referred to as the wide-angle head. All this work saw the power rise from 250bhp to 275bhp.

Mike Hawthorn was drafted into the Jaguar team for the 1955 Le Mans and, until the terrible accident, was involved in a magnificent duel with Fangio in the Mercedes, during which Hawthorn set a new lap record of 122.39mph. Hawthorn (and team-mate Ivor Bueb) won, albeit after the Mercedes team had withdrawn as a mark of respect.

The following year was one of mixed fortunes for Jaguar at Le Mans; all three factory entered D-types fell by the wayside, but honour was saved by the Ecurie Ecosse team, when drivers Flockhart and Sanderson took the chequered flag first. The Scottish team did it again the following year (when Jaguar as a company had officially withdrawn from racing), and privateers driving D-types also came second, fourth and sixth.

▲ *More power and beautiful flowing lines meant that the D-type could crack 170mph down the Mulsanne straight – a full 20mph faster than the C-type. The 9.0:1 compression engine mirrored the 1953 C-types in using three 45 DCO Weber carburettors (rather than the production SUs) and had still larger inlet valves, at $1\frac{7}{8}$in and a 'hotter' cam. This all added up to a power output of some 250bhp.* (Courtesy Jaguar-Daimler Heritage Trust)

SOMETHING SPECIAL

In 1957 Brian Lister started building his own sports cars powered by the already venerable XK engine. The Lister Jaguar soon became a force to be reckoned with and wins aplenty came the way of Archie Scott Brown, the Ecurie Ecosse team, Stirling Moss and Briggs Cunningham, to name but a few drivers. However, as others overhauled Lister in the development stakes, he withdrew from racing at the end of 1959.

▶ *The factory team entered in 1954 and the Rolt/Hamilton car was only just beaten (by less than a minute and a half) by the Ferrari of Gonzalez. The team entered again in 1955 and won, following the tragic events in this race, when an accident killed Mercedes driver Levegh and 81 spectators. For the following two years, it was the Ecurie Ecosse team who won with D-types – a happy hat trick commemorated on XK 150 badges. This is one of the Ecurie Ecosse D-types at speed around the famous Sarthe track on its way ...* (Courtesy Jaguar Cars Limited)

▶ *... to a 1956 victory. Early D-types used the 3.4-litre XK engine, but for later models the uprated 3.8-litre unit was fitted. The D-type was, of course, raced elsewhere, and where top speed was not so important, as at Le Mans, it could be used without the trademark tail fin. After this impressive winning streak, Jaguar enthusiasts were in for a long wait; though the company entered unsuccessfully in 1960 with the E2A (an E-type forerunner), it was 31 years before it once again conquered the 24-hour trial of speed and endurance, this time with the mighty V12, 700bhp XJR-9LM.* (Courtesy Jaguar Cars Limited)

▶ *Not that Jaguar didn't ponder an earlier return. Admittedly not with a 6-cylinder XK unit for propulsion, but who needs an excuse to include a photo of the sadly stillborn XJ13, 12-cylinder racer? The gorgeous flowing lines were, not surprisingly, the work of D-type designer, Malcolm Sayer. The dry-sumped 5-litre, dohc, V12 was reckoned to produce around 500bhp and is, unfortunately, probably most famous for being crashed at some speed following a tyre blow-out with Jaguar test driver, Norman Dewis at the wheel. Despite rolling several times, Dewis escaped unharmed and the car was able to be rebuilt. However, it was never fully developed and did not race, although how it would have performed against the contemporary Ford GT40s of the late 1960s is one of those tantalising questions to which we shall never know the answer.* (Courtesy Jaguar-Daimler Heritage Trust)

▶ *In modern classic racing, Brian Talbot is having fun in his D-type replica. This stunning machine looks, goes and sounds like the real thing – it's just cheaper if you bend it!* (Courtesy Rob Beere Racing Services)

E-TYPE

A prototype E-type (clearly developed from the D-type, but with the benefit of independent rear suspension) was called the E2A and was entered in Le Mans by Briggs Cunningham in 1960. It used a 2997cc 6-cylinder engine with an alloy cylinder block, which produced 295bhp at 6,800rpm. Its D-type predecessor needed a full 3781cc to achieve the same output. However, its reliability was not in the same league as the (then) current XK engines (the 3.4 and 3.8). Briggs had a 3.8-litre XK engine installed (which necessitated the bonnet bulge which became part and parcel of all official E-types) and, though it was raced in anger, it couldn't match the success of its predecessors.

After the introduction of the E-type road car there was a racing debut that many manufacturers would kill for; two cars came first and second in the GT Trophy race at Oulton park – driven by Graham Hill and Roy Salvadori. When one considers they were lining up on the grid against such heavyweight opposition as the 250 Ferrari and the Aston Martin DB4, that's no mean feat!

Success at Le Mans, however, was not to come the way of the E-type, not least because factory-financed opposition was just too much to cope with by under-funded privately run teams. That said, E-types came fourth and fifth in the 1962 event, although this was more because they didn't break down than through any particular turn of speed.

Probably the most famous racing E-type was the Lightweight (officially, the Competition E-type). As the name suggests, the aim was to get the weight down, something achieved by making everything possible (body panels, etc) from aluminium rather than steel. It produced around 300bhp from its dry-sumped, XK engine which featured Lucas fuel injection rather than SU or Weber carburettors, and an alloy block to match the D-type, wide angle cylinder head. The expense of producing the car was astronomical, not least since just 12 were manufactured, and the cost of recreating your own replica requires a Cray computer to calculate the final invoice!

▲ *Modern production racing championships always see their fair share of E-types, mixing here at Hockenheim with offerings from Ferrari, Porsche, Lotus and TVR, and* ... (Courtesy Rob Beere Racing Services)

▲ *... as if that wasn't enough, they also compete in the Anglo/American series. This is Michael Evans pushing his E-type hard to stay ahead of a TR6, a Corvette and a brace of Lotuses (Loti?).* (Courtesy Rob Beere Racing Services)

SALOONS

It's tempting not to think of the XK engine competing under the bonnet of a saloon car, but it's a rather large, luxury Jaguar limousine that holds the honour of being the only XK-engined car to win the Monte Carlo Rally. In 1953 the Appleyards were forced to switch from the XK 120 to the Mk VII saloon (because of rule changes) and won the Alpine Rally once more and finished second in the Monte Carlo. Stirling Moss won the Silverstone International Trophy in a Mk VII, and in 1954 it was Appleyard who took the honours, beating Rolt and Moss into second and third places. In the 1954 Monte Carlo, Mk VIIs took 6th and 8th places. In 1955 it came 4th overall (with a class win) in the Tulip Rally and won its class in the RAC Rally, but it was the following year that true glory and that Monte Carlo win came the way of the Mk VII.

The lighter Mk 1 and Mk 2 cars were very successful in various disciplines. Paul Frère (more usually associated with Porsches) won at the Spa Production Car race in 1956 driving a Mk 1 2.4-litre, and for many the Mk 1 3.4 saloon will always be associated with Mike Hawthorn, a Jaguar works driver in 1955 and 1956. He drove it to its inaugural victory at the *Daily Express* race in 1957, the first of many for the popular British driver. It is a sad irony that when he was killed in a road accident in 1959, he was driving his own 3.4 saloon.

The Mk 1 and Mk 2 models were used almost universally in saloon car racing and won more times than can be counted in the hands of a whole galaxy of motor racing names, such as Stirling Moss, Mike Hawthorn, Jack Sears, Graham Hill, Colin Chapman (founder of Lotus), Danny Hulme and Bruce McLaren.

By 1961 Jaguar had all but taken over the Silverstone production car race, having won for the tenth year in a row – convincingly, as it happens, with Mk 2 3.8S models taking the first five places!

▼ *Classic saloon championships see more than a few XK-powered Jaguars. Leading in this shot is Martin Sutherland's 4.2-litre XJ6, with a 420 hot in pursuit.* (Courtesy Rob Beere Racing Services)

▲ *There's no escape in the Jaguar Car Club's Bridgestone Jaguar Racing series – a real feast for the Jag enthusiast.* (Courtesy Rob Beere Racing Services)

Chapter 5

The Impressionists – XK Kit Cars

It's not surprising to learn that such a prodigious and technically adept engine as the 6-cylinder XK unit has been used for many kit car applications. In most cases the purchaser has the option to buy the new car fully built, complete with engine and running gear, or as a kit in various stages, and to add his own parts as required. Clearly, if you have a Jaguar which has terminal body rot but good engine, transmission, suspension, etc, building your own sports car is not short of appeal. Here are just a few of those currently on the market.

SUFFOLK SS100

◀ The basic car comes with a 4.2-litre XK engine fitted with twin SU carburettors and producing 182bhp. Other engine/gearbox combinations are available on special order. The front and rear suspension, steering and 4-speed gearbox are also Jaguar XJ6-sourced. The bodywork is GRP (glass reinforced plastic) and covers a tough ladder chassis which is anti-rust warranted for 10 years in the UK. Inside, the driver is cosseted with Wilton carpet, Connolly leather and a duck down filled folding hood.

The chassis is different, but just about everything else looks familiar. The XK engine is mated to a Jag gearbox, drivetrain and suspension.

◀ The hand-crafted and louvred bonnet folds back to reveal the heart of the matter, a gleaming 4.2-litre engine good enough to eat.

▲ *Is it real or is it a replica? Suffolk is keen to make the point that this car is recognised by the JDC as a Historic Replicar, and no wonder!*

ROYALE DROPHEAD

▶ *Although also using the XK engine and some Jaguar parts, the Royale is most definitely unlike anything Browns Lane ever produced. With the look and feel of a 1930s open-top tourer, the Bugatti-like Royale, takes most of its mechanical bits and pieces from a donor XJ6, including the 4.2-litre engine, 3-speed gearbox, suspension and power-assisted steering. A material called Coremat is sandwiched between the layers of GRP to give the body shell extra strength and rigidity. Most of the brightwork is either chrome on brass or stainless steel, and wood, leather and Wilton are predominant inside the car.*

The Royale brings back everything from the luxurious age of 1930s long-distance tourers – except the inherent unreliability! Being XJ6-based, spares are available virtually anywhere.

BROOMSTICK BS120

▶ *Fancy an XK 120? Of course you do, but for most of us the price of a pristine car would be a tad prohibitive. This is where the BS120 comes in, offering XK 120 style and performance for Mondeo money – truly magic! Based on a ladder chassis, the BS120 uses parts from the XJ6 Series cars (1, 2 and 3) and can be fitted with either the 2.8, 3.4 or 4.2-litre engines. So, take one immensely strong ladder chassis …*

◀ *… add this many GRP body panels, add XJ6 running gear with an XK engine and …*

▲ … *your very own Jaguar sports car for considerably less than the cost of the real thing. As with most modern replicas, it's an incredible likeness.*

ARISTOCAT

▶ *Autotune reckons that the Aristocat is not a replica, but a car within the spirit of the whole XK series. As ever, a complex ladder chassis provides the basis for the GRP body and mechanical parts from a donor car – the best bet being a manual XJ6. All three Jaguar clubs have accepted the Aristocat as legitimate. There are two styles available, this being the drophead version with …*

▲ *... a coupé also on offer. The similarity between them and the original Jaguar XK cars is not surprising, as the dimensions are almost identical.*

LE MANS COMPANY

▲ *Richard Purser runs the Le Mans Company and produces this glorious looking D-type. A superb-looking kit car, the doors are pre-hung on the GRP body and the front and rear sections are pre-bounded with chassis mounting lugs. It also comes complete with all electrical components (lights, loom, etc). You can choose between a Perspex deflector or full windscreen and weather gear, depending on how true to the era (and how much you like the UK weather!) you want to be.*

▶ *The Chief Engineer is Bob Needham, who served with Browns Lane from the mid-1950s, so he knows his way around the XK engine better than most. Though you can install your own engine in the D-type (or, heresy, even a Nissan straight six!) you can also opt for tuned versions from Le Mans like this beauty. It's a 4.2-litre model with 9.5:1 compression ratio. It breathes through triple 45DCOE Webers and produces around 290bhp. The valves are 1⅞in and 1⅝in for inlet and exhaust respectively, short skirt pistons are used and the head has been polished and ported by VSE. The cam is much hairier than standard and has a profile based on that of the original D-type.*

▶ *The stainless steel headers were supplied by Falcon, but they needed to be modified by Le Mans to suit this specific application. It's a shame to put this engine under a bonnet, really.*

On the cards at the time of going to press is a version based on the D-type-inspired XK SS model. Can't be bad!

PART 2

Practicalities

Chapter 6

Workshop Procedures and Safety First

Working on your XK engine, whether it's simple maintenance and repair or a complete rebuild, will give you immense satisfaction. However, it's important to take a leaf or two from the professional mechanic's book of safety. A logical approach to DIY motoring is essential, as are the correct procedures and the use of the right tools for the job.

PREPARATIONS

When you're working on your car, always make sure that someone knows where you are, and ask them to check up on you on a regular basis – it could be your life-saver.

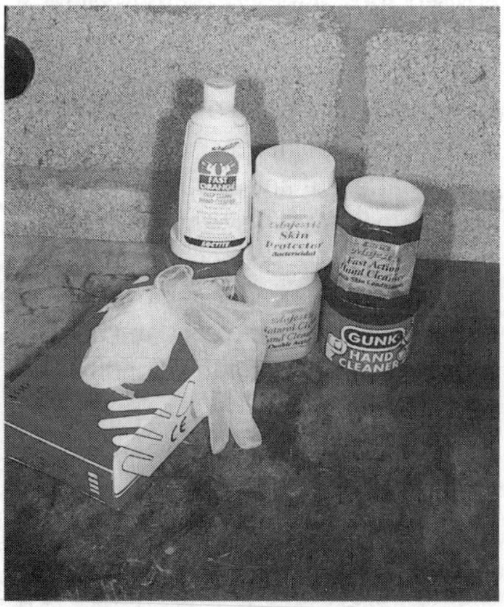

◀ *It's always wise to protect your hands, not least from the unpleasant effects of oil and grease which are now known to be carcinogenic. Use a quality barrier cream before you dirty your hands so that any dirt on them will simply wash off. Better still, wear vinyl or latex gloves to keep dirt and oil off your skin altogether. When you wash your hands, use a purpose-made cleaner which will pull off the dirt without removing the skin's natural oils.*

- Wear overalls, not least because there are less likely to be loose items of clothing to get caught in moving parts.
- Impose a NO SMOKING rule in your workshop at all times.
- Remove watches and jewellery where possible.
- Keep things tidy – a workshop with cables and pipes running all over the floor is a recipe for disaster.
- Read the instructions! It's best to do this in the warmth of your living room, as you'll be more likely to skip bits in a cold workshop.
- When cleaning components DON'T use petrol (gasoline). Use white (mineral) spirit, paraffin (kerosene) or purpose-made industrial cleaners.
- When using any power tool, especially a grinder – and even when using hand tools where there is a danger of flying fragments (chiselling a seized nut, for example) – wear goggles to protect your eyes.

THE CAMERA NEVER LIES

If you're working on particularly complex items, make notes and diagrams as to where the various component parts should go. Alternatively, use a camera or even a video camera to show a dismantling procedure (the replacement order in reverse). Using a video camera also gives you the opportunity to add a 'commentary'. Don't believe that you'll remember everything – it's amazing how much the memory fades even after just a week, and many restoration jobs will take months, maybe years.

UNDER THE BONNET

- Make sure that the gearbox is in neutral, or 'Park' in automatic gearboxes, before starting the engine.
- Don't leave the key in the ignition while you're working on the car.
- Always remove the coolant filler cap with a cold engine. If you *have* to do it when the engine is warm, do it slowly, with a cloth for protection. Undo slightly and let the cap settle at the first indentation. This will release the pressure and steam – but remember that the steam will be even hotter than the escaping coolant.

- Keep brake fluid and anti-freeze off your paintwork. Wipe up spillages straight away.
- Never siphon fluids such as anti-freeze or petrol (gasoline) by mouth – contact with skin as well as inhalation can damage your health. Use a suitable hand pump and wear gloves.

GENERAL SAFETY NOTES

- Take care not to inhale any dust, especially brake dust, which may contain asbestos.
- Wipe up oil or grease spillages straight away. Use oil granules (cat litter will do the job!) to soak up major spills.
- Use quality tools – an ill-fitting spanner could cause damage to the component, your car and, of course, to yourself!
- When lifting heavy items, remember the rule; bend your legs and keep your back straight. Know your limitations – if something is too heavy, call in a helper.
- Time is a vital element in any workshop. Make sure you've got enough to finish a job. Rushed work is rarely done right.
- Children are naturally inquisitive but don't allow them to wander unsupervised around or in your car, especially if it is jacked up.

FINISHING

When you've finished, clean up the workshop and clean and replace all your tools. You'll reap the rewards in time saved next time out, and in tools that last much longer and work much better.

FIRE

Always carry a fire extinguisher in the car and have another one available for use in your workshop. Choose a carbon dioxide type or, better still, dry powder, but never a water type extinguisher for workshop use. Water conducts electricity and, in some circumstances, could actually make an oil- or petrol-based fire worse. Check the instructions before you need to use the extinguisher and always direct the jet at the base of the flames.

Naked flames (when using gas welding equipment, for example) are an obvious threat, but petrol and its vapour can be ignited even by a spark – which could be caused by something as simple as a short circuit. This is why you should always disconnect the battery earth terminal before starting work. Other problem areas include sparks from a grinder, the striking of two metal surfaces against each other, the bulb in an inspection lamp or even a central heating boiler starting up.

MAINS ELECTRICITY

Mains power tools should be used with care outdoors. Use the correct type of plug with correct and tightly-made connections. Where applicable, make sure that they are earthed (grounded).

Use an RCD – a residual circuit breaker which in the event of a short circuit cuts the power immediately to reduce the risk of electrocution. RCDs can be purchased quite cheaply from most DIY stores. Take special care when working in damp conditions, especially if you are using a mains extension lead. Wherever possible, work indoors and/or use battery powered tools.

FUMES

Never run your Jaguar in an enclosed space – the exhaust fumes contain deadly carbon monoxide and can kill within minutes. Treat all chemicals (not least of which, petrol) with great care. Many cleaning agents and solvents contain highly toxic chemicals and should not be used in confined spaces or be used for long periods without a break. Wear gloves when working with chemicals and if any is spilled on your skin, immediately rinse off with water.

HIGH TENSION IGNITION

Touching parts of the ignition system with the engine running (or being turned over), notably the HT (high tension) leads themselves, can lead to severe electric shock, especially if the vehicle is fitted with electronic ignition. Voltages produced by electronic ignition systems are much higher than normal and could prove fatal, especially to those with pacemakers.

The likelihood of an electric shock is greater in wet or damp conditions, when a spark can 'jump' to an earthing point – you! Take great care when performing a task which requires the engine to be running (setting the timing, for example) not to touch any ignition components.

PLASTIC MATERIALS

The use of fibreglass and other similar 'plastic' materials is opening up a whole new area of safety awareness. It is vital that the instructions be followed to the letter to avoid a dangerous situation developing. Substances such as polymers, resins and adhesives produce dangerous fumes (both poisonous and flammable) and can be skin irritants. In particular, do not allow resin or 2-pack hardener to come into contact with the skin. Make sure such materials are clearly labelled and stored safely away from the reach of children and/or under lock and key.

THE BATTERY

Most batteries give off a small amount of highly explosive hydrogen gas, so you should never allow a naked flame or a spark near your battery. Always disconnect the battery earth terminal whenever you are working on your car to remove the possibility of an accidental electrical short circuit. When charging the battery, disconnect both negative and positive leads and, unless otherwise advised, loosen the filler caps to allow excess gasses to escape.

The battery electrolyte level should be kept topped up (using only distilled water) to the point specified on the side of the battery. If any electrolyte is spilled, wipe it up immediately and wash off any that gets on your skin – it is highly corrosive. If you need to remove your battery, wear rubber gloves and goggles, always keeping the battery upright.

PETROL SAFETY

Petrol is a highly flammable, volatile liquid and should be treated with great respect. Even its vapour will ignite at the slightest provocation. When not actually in your fuel tank, it should be kept in metal cans (or approved 'plastic' cans) and stored where there is no danger of naked flames or sparks. Cans should have a ventilation hole to prevent the build-up of vapour. If you work in a pit, extra care is required, as petrol vapour is heavier than air and will tend to build-up in the bottom of the pit.

ENGINE OILS

There is some danger from contaminates that are contained in all used oil and, according to some experts, prolonged skin exposure can lead to serious skin disorders. You can offset this by always using barrier cream on your hands and wearing plastic or rubber gloves when draining the oil from your engine or transmission.

OIL DISPOSAL

Never pour used oil down a drain or on to the ground. It is environmentally unfriendly and renders you liable to action from your local council. In most EC countries, including the UK, local authorities must provide free Oil Banks as a safe means of oil disposal. If you're unsure where yours is, contact your local Environmental Health Department for advice or ring the Oil Bank Line on freephone 0800 663366 for details of your nearest bank.

FLUOROELASTOMERS

Many items found on modern cars (eg oil seals, gaskets, diaphragms, and O-rings) may appear to be rubber but in fact are made from synthetic substitutes which contain fluorine. The materials are called fluoroelastomers, and if heated to more than 315°C (600°F), they can decompose in a dangerous manner. Indeed, some decomposition can occur at temperatures of around 200°C (392°F). These temperatures would normally only be reached if a car were to be set alight or if it were 'broken' by a vehicle dismantler using a cutting torch.

Where there is any water present, including atmospheric moisture, the heated fluoroelastomers produce extremely dangerous by-products.

The Health and Safety Executive says: 'Skin contact with this liquid or decomposition residues can cause painful and penetrating burns. Permanent irreversible skin and tissue damage can occur.'

Clearly, this is important to note if your car has caught fire, even if only partially, or if it has been stolen and 'fired' by the thieves. Even more caution is required if you are searching for used parts at a vehicle dismantlers.

▲ *To save transporting old oil five litres at a time, use a large drum (say 25 gallons) as interim storage. When it is full, take it for safe disposal.*

▲ *Always have a fire extinguisher to hand and know how to use it. There are so many things in your workshop and on your Jaguar that are flammable, you can't afford to be without one. Ideally, buy a smaller extinguisher to carry inside your vehicle, and a larger model for workshop use.*

Observe the following safety procedures:

1. Never touch blackened or charred pieces of rubber or anything that looks like it.
2. Allow all burnt or decomposed fluoroelastomer materials to cool down before inspection, investigations, tear-down or removal.
3. Ideally, don't handle parts containing decomposed fluoroelastomers. If you have to, wear goggles and PVC protective gloves whilst doing so. Never handle them unless they are completely cool.
4. Contaminated parts, residues, materials and clothing – including protective clothing and gloves – should be disposed of by an approved contractor to landfill or by incineration according to national or local regulations. Oil seals, gaskets and O-rings, along with contaminated material, must not be burned locally.

▲ *The gap between removing an engine for rebuilding (or even starting work on an engine whilst it is still in the car) can often be weeks, or even months. Also, many Jaguars are used as weekend and/or fine-weather cars. This means that the battery gets little or no use. Apart from the fact that a flat battery won't turn the engine, a battery that is allowed to fully discharge on a regular basis is going to wear out much more quickly than it should. Moreover, batteries left uncharged during the winter months can freeze and crack their casings. Worse still, a battery allowed to reach a condition of deep discharge may never accept its full charge again.*

Remembering to go out and start the car on a regular basis will help the battery, but running the engine at idle on the choke will do the engine a power of no good. Enter the AG Products Battery Sava, a simple, effective and well-priced product aimed at saving your battery whilst the car is not being used. Looking like a calculator transformer, it clips on to the battery terminals (with the car's leads still in place) and then plugs into the mains. According to the makers, it costs a mere 4p per week in electricity. There are two versions – above and below 45AH batteries, to cater for the motorcycle and car markets respectively.

GET OUT AND GET UNDER – RAISING YOUR JAGUAR SAFELY

There are times when you'll need to raise your Jaguar for engine work, from the extreme of removing the engine completely or even just for better access to the starter motor or radiator hoses. A trolley jack, axle stands and ramps must be seen as essential tools. Don't skimp on quality when you're buying, or on the safety aspects when you're using. According to the latest statistics, 20 people die every year from being crushed by collapsing cars – don't get your name on that grim list.

- Never work under your Jaguar when it is supported only by a jack. Use additional support from securely placed axle stands.
- Protect your eyes and hands using goggles and gloves.
- Don't attempt to loosen high-torque nuts or bolts while the car is off the ground. Torque them up when the vehicle is back on the floor.
- Never touch any part of the exhaust, manifolds or catalytic converter before ascertaining whether or not they are hot.
- Catalytic converters get extremely hot – don't park over dry grass, oily rags or

▲ *A trolley jack, axle stands and ramps should be seen as essential Jaguar restoration and maintenance tools. Don't skimp on quality when you're buying lifting equipment or on the safety aspects when you're using it.*

▶ *When working underneath your Jaguar, the safest way to raise it is to use ramps. Always enlist the aid of a helper to ensure that you drive up the ramps squarely and with a wheel in the centre of each ramp. Drive up to the end stops on the ramps, but not beyond. Chock the wheels remaining on the ground. Note the strips of carpet wound round the end of the ramps which help the wheels get a purchase on the rungs and help prevent the ramps skidding.*

any other material which may catch fire from the heat. Never run a catalyst-equipped engine without the heat shield in place.

● Although engine and transmission fluids should be warm in order to drain them properly, ensure that they are not hot enough to scald.

◀ *This is SIP's Omega 3-ton Pro jack. Clearly, its SWL (safe working load) gives plenty of leeway, and it has a large (140mm/5½in) saddle, so there's little danger of it slipping. Where you are jacking on parts that might be damaged superficially, you can use the rubber protective cover provided, and its small extra 'Fast Lift' handle is very useful in getting the jack arm to the jacking point quickly. It has a min/max lift height of 133mm/483mm giving a lift span of 350mm. Being of professional standard, the long main handle turns cog gears to release the pressure, thus obviating the need for constantly removing the handle to turn the valve manually.*

▲ *Having removed a wheel it's a good idea to put it under the car as a 'belt and braces' measure.*

● If you're removing the wheels, slacken off the wheel nuts before the car has been raised.
● Torque up the nuts when the wheels are on the ground.
● Never use ramps, jacks or axle stands on a 'soft' surface, such as tarmac.

SCISSOR JACKS

If your car has a scissor jack as its original equipment, it should only be used for emergency wheel changing or, at most, raising the car prior to supporting it on axle stands. Only use the jack at the specific jacking points or certain strong parts of the vehicle – to do so elsewhere could be extremely dangerous, not to say expensive. If you're not sure where they are, check your handbook or *Haynes Workshop Manual*. In many cases the jacking points will be rotten, and great care should be taken in case the car slips whilst partially raised.

USING A TROLLEY JACK

Leave the vehicle out of gear and with the handbrake off until you have reached the required height. This allows the vehicle to move as the jack 'rolls' into place. Exercise **extreme caution** during this procedure and ensure that no-one can be hurt if the car suddenly slips off the jack. Re-apply the handbrake, put the vehicle in gear and chock the wheels remaining on the ground.

Always use axle stands as well as the jack to support the weight of the vehicle. Once the vehicle is safely at the right height, remove the ignition keys. If you need to run the engine for any reason, make sure you put the transmission in neutral.

IMPORTANT NOTE: Always use a strong piece of wood on the jack saddle to prevent damage when jacking your Jaguar directly on to the crossmembers, etc.

Chapter 7

Tools and Techniques

Using quality tools is one of the cornerstones of successful maintenance and restoration. Good quality tools are seldom cheap but, overall, they are often worth far more than their purchase price. To be able to complete a job in minutes with the right tools, rather than having to spend hours with the wrong ones (especially during a winter restoration job) makes them worth their weight in gold.

'Sunday market specials' are often manufacturers' rejects and likely to break or fail under pressure. A good warranty is a sign that manufacturers have faith in their products, and almost all good tools will be guaranteed for a lifetime of normal use. Something to check is that each tool (or even individual parts, such as sockets) is marked with the manufacturer's name – if not, how can you make a claim if it proves to be faulty?).

▶ *Hurling your tools, dirty and oil stained, in a heap in the corner of your garage is a recipe for trouble. The next time you need them, you'll have to spend precious time searching and, if they've not been cleaned, they will gradually function less and less efficiently. So, after the rigours of whatever job you've just completed, wind down with a little therapeutic tool care. Wipe them over with a rag and WD40 (which is a great cleaner) before putting them back in their places in your tool chest. This way they'll last much longer and give better service. Remember the adage – look after your tools, and they'll look after you.*

SPECIAL TOOLS

Where a tool is required for a specific job but will not be used again for years, it will often make no financial sense to buy it. This is where an owners' club can help, with most having a tool-hire facility of the most common 'one-off' items. Hiring the tool(s) you require for, say, a week, will allow you to do the job properly, and without breaking the bank. Moreover, it will remove the temptation to 'make do' with incorrect tools which, at best, could lead to vehicle or component damage and, at worst, to personal injury.

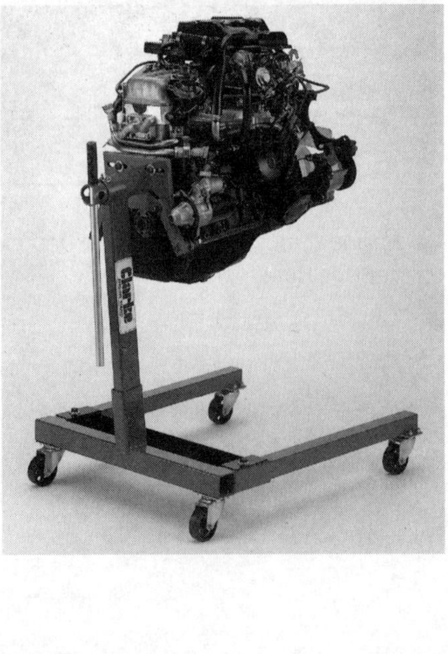

▶ *Essential for serious engine-out work and/or rebuilds is an engine stand. Clarke's CES750 model can handle a unit weighing up to 340kg (750lb) and, once mounted, the engine can easily be turned through 360 degrees. There is a smaller model, the CES500, rated at 227kg (500lb).* (Courtesy Clarke International Limited)

▲ *To get the engine out in the first place, you'll need a heavy-duty engine crane. Quality is important here, as it is usual to remove the XK engine and gearbox as a single unit (combined weight 400kg plus). Clarke's Strong-Arm range comes in a variety of capacities and lifting heights, ranging from 500kg to a full 2 tonnes. The folding type is particularly useful for the DIY mechanic, as it takes up far less space than conventional cranes.*

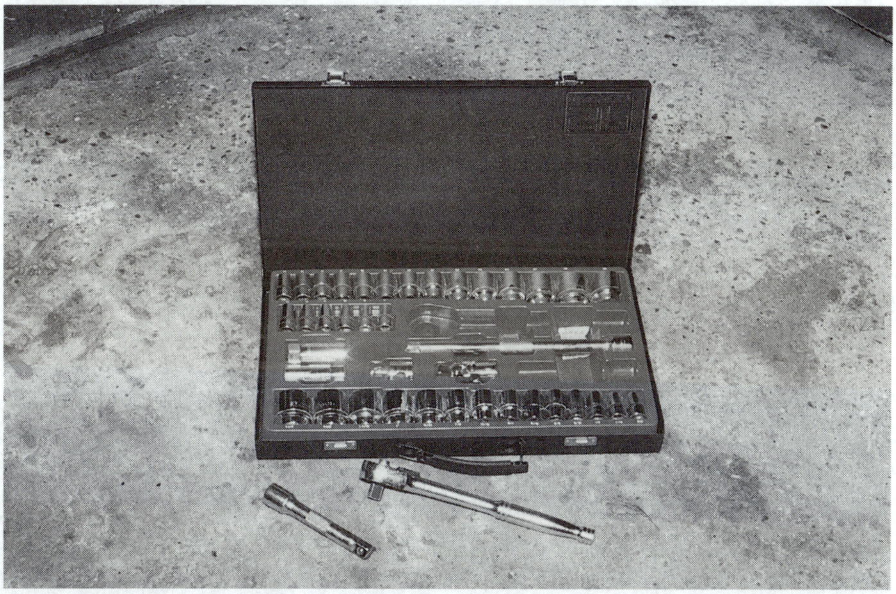

▲ *A socket set is an essential part of restoration and maintenance. A 3/8in drive set is generally suitable for most applications, but for some fasteners, you'll need a 1/2in drive to cope with the torque. The Jaguar is a mix of imperial and metric sizes (and some equivalents) so it pays to have a selection of both – as in this steel-boxed Sykes-Pickavant set. Take note of the various types of socket now available. Many companies offer bi-hex (ie 12-sided) sockets, which have the advantage that they can be fitted on a fastener in twice as many positions – sometimes a consideration where space is tight. However, for high torque applications, it's generally best to opt for a conventional hex (6-sided) socket, where the twisting power is being transmitted through a larger area.*

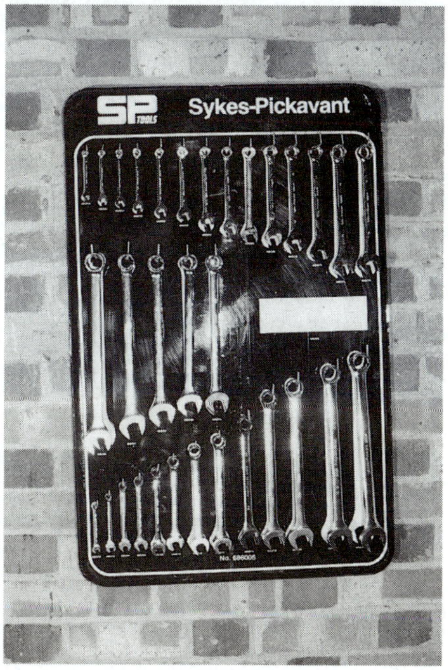

▲ *There are three basic types of spanner; open-ended, ring and combination. A combination spanner set is particularly useful and probably the best place to start. There are open-ended and ring variants of the same size on the same spanner. The Sykes-Pickavant mounting board shows silhouettes of each spanner, which enables the user to see at-a-glance if any are missing.*

SPANNER JAW GAP COMPARISONS

Actual size	*Nearest metric size*	*Metric size in inches*	*AF size (mm)*
0.248	7	0.276	4BA
0.320	8	0.315	2BA
0.440	11	0.413	7/16 in
0.500	13	0.510	1/2 in
0.560	14	0.550	9/16 in
0.630	16	0.630	5/8 in
0.690	18	0.710	11/16 in
0.760	19	0.750	3/4 in
0.820	21	0.830	13/16 in
0.880	22	0.870	7/8 in
0.940	24	0.945	15/16 in
1.000	26	1.020	1 in

IMPORTANT: A torque wrench should *never* be used as a 'breaker bar' for loosening particularly tight fasteners.

▶ *It can be dangerous (and not a little expensive in terms of broken engines) to over- or under-tighten nuts and bolts, and for this reason you'll see torque figures given throughout this book. Use a quality torque wrench to ensure this doesn't happen. Seen here are two S-P Motorq wrenches, ⅜in and ½in drive, which come with the all-important hard plastic cases – a torque wrench is a precision instrument and should be treated as such. Twisting the end of the wrench locks the device so that accidental changes cannot be made. To avoid placing too much strain on the wrench, when you've finished using it, wind the adjuster right back to its lowest setting, but NOT below it.*

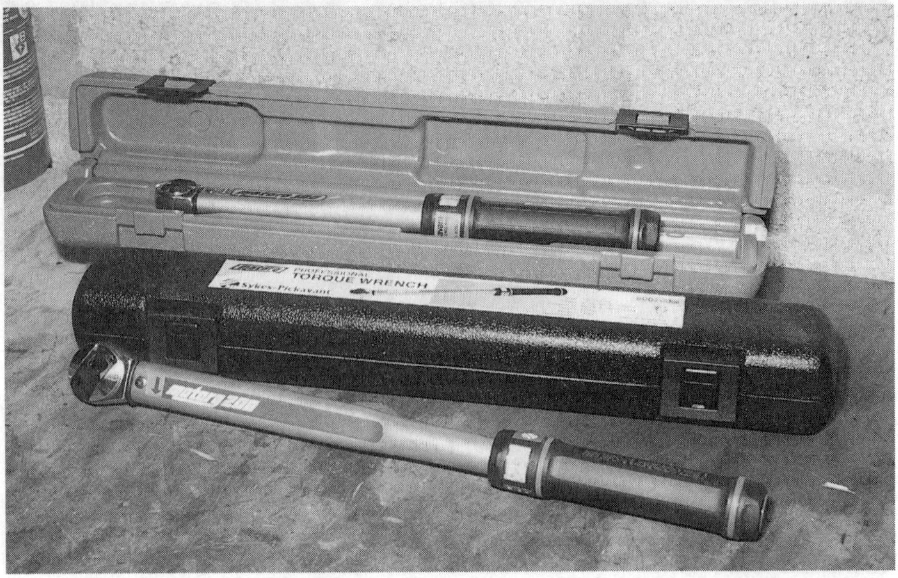

▶ *Buying screwdrivers in a set like this is one way to ensure you've always got the right 'driver for the job. Using the wrong one is a recipe for disaster. Remember that there's a definite difference between Pozidriv and Philips head screws. On most cars made after 1970, the screws will be Pozidriv. If there's absolutely no choice, a Philips screwdriver will just about fit a Pozidriv head but NOT the other way round.*

▶ *Some 'drivers, like this one from Teng, have hex head shanks, meaning that a spanner (preferably a ring, as here) can be used to apply the turning force. Note also the hole in the end of the handle, designed to accept a tommy bar, again enabling more turning force to be applied.*

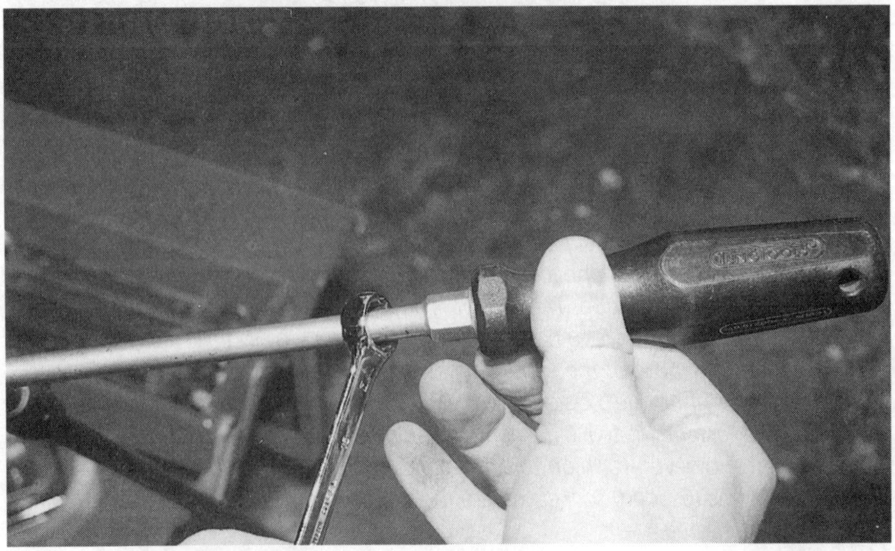

▲ *Like screwdrivers, it's important to use the right pliers for the task in hand. A standard pair and long-nose pair should be seen as a minimum, though for specialised use such as circlip removal, you'll need circlip pliers. This particular S-P kit includes a range of 'jaws' for different sizes of internal and external circlips.*

▲ *Almost all aspects of Jaguar engine restoration and maintenance will require the use of a puller at some point. It is possible to buy individual pullers but S-P's Multi-puller Pack saves money in the long run by combining in different ways to make five different types of puller.*

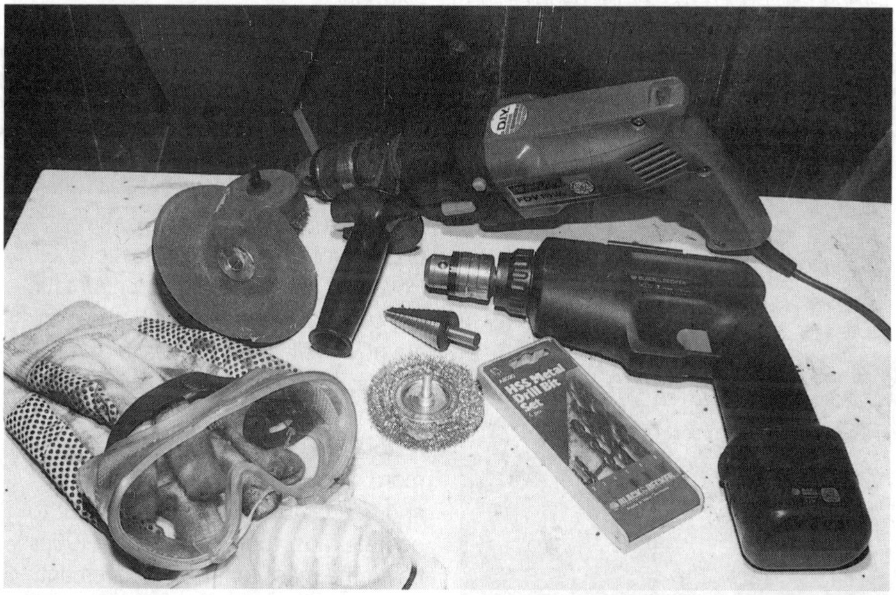

▲ *The use of power tools will make your Jaguar life much easier, especially if you don't have access to air tools (see later). A mains-powered drill should be seen as a first step, with a cordless model close behind. Each can be used with a wire brush to clean up rusty surfaces.*

POWER TOOLS

SAFETY: Always take great care when using power tools. Read the notes about safety and mains electricity and use goggles and gloves when using drills and wire brushes, etc. If using a mains-powered device outside with an extension lead, make sure no-one can trip over it and that it is not vulnerable to rain water or moisture in any way.

◀ *A bench grinder is a useful tool. This 6in disc model is one of Clarke's range of 240v bench-mounted grinders and comes with spark arresters, tool rests and a dust extractor on each wheel. All have eye shields which should ALWAYS be in place when you're working. Many such machines have one grinding wheel and one buffing wheel, thus doubling the usefulness.* (Courtesy Clarke International Limited)

HEATING

SAFETY: Always follow the safety rules included with any new heater.

▶ *A major problem with working on your XK engine (in the UK at least) is heat – or, more usually, the lack of it. Unless you've got a very understanding family who will loan you the (very strong) kitchen table for a few months, you'll be outside in the garage or workshop. A garage attached to a house will generally be warmer, but for many months of the year, the ambient temperatures are likely to be near the lower end of the thermometer. Apart from the unpleasantness of being cold, there's the real danger that jobs are going to be rushed, and rushing is seldom a prerequisite of doing something properly. The SIP range of propane Fireball heaters is designed to suit most garage sizes. The Fireball 40 produces 10kW of heat and is reckoned suitable for a space of around 8,000cu ft. The mid-range Fireball 80 model produces 23kW of heat and is designed for spaces of around 20,000cu ft. This is the mighty Fireball 120, capable of pushing out 35kW of heat and designed for larger garages and workshops (around 30,000cu ft) – or for people who feel the cold! All Fireball models feature a turbo fan which gives maximum heat spread and efficiency. All are virtually maintenance-free and come complete with hose, regulator and a 12-month warranty.*

AIR COMPRESSORS

Once upon a time, the use of a compressor and air tools was the preserve of the garage professional. However, the cost of going pneumatic has come down dramatically and prices are now well within the reach of most DIY motorists, all of whom will benefit. Anyone contemplating a full or part-restoration project will benefit even more. An air pressure of between 3.4 and 6.8 bar (50–100psi) is sufficient to operate most tools, with 6.2 bar (90psi) being the most common requirement. Lubricate all pneumatic tools on a regular

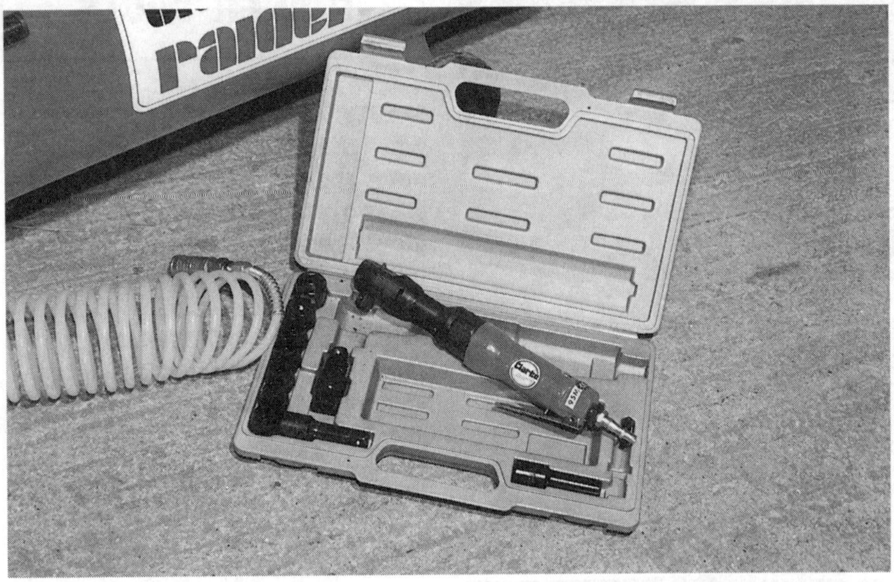

▲ *Choose your compressor carefully. Clarke have a vast range of compressors from small DIY units to heavy duty industrial models. Though it may be tempting to go for the cheapest, think about future possibilities to avoid being 'penny wise and pound foolish'. The Raider Air 50 is a very popular mid-range model and ideal for most aspects of Jaguar restoration. Its 3hp motor produces 150psi and it has a large (50 litre) tank, meaning that the motor is not always cutting in and out. In terms of safety, it has a hefty thermal overload with reset. Whereas using three electric tools would require three cables (and probably an extension lead), using three air tools requires just the one hose. Note here, the 'curly' air lead, probably the best bet for the cluttered DIY garage, where it will coil back on itself out of the way when not being used. A handy feature of this particular model is the twin air outlet tap, allowing two tools to be connected (though not necessarily used) at the same time. Once you have the compressor, tools can be added almost ad infinitum. For engine work, start with a ratchet set and, to be used with some discretion, an impact wrench.*

basis by squirting a few drops of oil either down the air connection or into a special oil hole. Air contains moisture and this will collect inside the mechanism and cause rust. Oil daily or before every use.

SAFETY: Always follow the safety guidelines supplied with any compressor or tools. Wear protective gear such as gloves and/or goggles where appropriate.

▶ *The Sykes-Pickavant blow gun is cheap and incredibly useful for cleaning off loose dirt from the side of the engine, drying off parts that have been cleaned using solvents, blowing through oilways and cleaning out carburettors, etc. (Note that goggles should always be worn when using such tools.)*

Chapter 8

Engine Notes

The XK engine changed in innumerable ways over the 40 years of its production, but in essence the five production 6-cylinder engine capacities all share the same basic layout. The blocks are cast iron and the cylinder heads are aluminium with crossflow hemispherical chambers, housing double overhead camshafts driven by two Reynolds duplex roller chains, and at the bottom end of the engines is a 7-bearing crankshaft, built for longevity and silky smooth power delivery.

Essentially, the engine is immensely strong, powerful and torquey and it is capable of covering huge distances between rebuilds – if it is treated correctly. However, it's all too common to find that even regular maintenance and servicing has been skimped or ignored altogether. Though it may appear complex, even something as simple as using a high-quality engine oil and changing it on a regular basis can add many thousands of miles to the engine's useful life.

The XK unit shares a characteristic with that other long-lived engine, the Rover V8, in that it is prone to sludging up internally if the oil is not changed regularly. A build-up of sludge around the crank prevents the bearings getting enough oil, the ultimate expensive and needless result of which is an engine rebuild. It is also thought to be one of the contributory factors to weeping oil around the rear main bearing. The absolute oil-change limit is 5,000 miles, but on cars used infrequently it should be more often – 3,000 miles or less. Good quality oil may seem expensive, but not compared to a full rebuild!

A characteristic of many pre-1970 XK engines was a puff of smoke from the exhaust as the engine was started. Not unnaturally, this was seen by those who know as proof positive that all XKs are oil-burners and hence big trouble. But it ain't necessarily so, for some XKs of that vintage simply didn't have inlet valve oil seals; meaning that when the engine had stood for some time, the oil would drain down them. Obviously, on start-up, this oil would be blown out of the exhaust in a puff of blue smoke. It's common to find that, where these are not fitted, professional engine rebuilders add them as a matter of course – an option also open to the DIY builder.

The 3.8-litre XK engine is still regarded by many enthusiasts as the best of the lot, although there are probably just as many that can produce a reasonable argument to the contrary. Whatever, it's an impressive engine with the potential to produce plenty of smooth power and torque, which is what Jaguar motoring is all about. Initially, though, its oil consumption was disappointing, to say the least – many owners reckoned they were lucky to get 400 miles per pint. The main reason for this was that the bores tended to flex and, in doing so, allowed oil past the piston rings up into the combustion chamber where it was burned. It was 1964 before Jaguar took steps to rectify the problem by fitting different pistons. These were chamfered and had oil drain holes below the control ring.

The short-lived 2.8-litre engine, as fitted to Series 1 XJ6 models has never been popular, not least because it's really too little engine for too much car. Add a reputation for unreliability and you've got enough reasons to choose another engine.

When the 4.2-litre engine was introduced (effectively replacing the 3.8-litre version) Jaguar was keen to retain the original exterior block dimensions. This kept down development and production costs and meant that it would physically fit where the smaller engine had already been installed. Four of the cylinder bores changed position; the two end bores (1 and 6) were moved out slightly and numbers 3 and 4 were squeezed in a little. Numbers 2 and 5 stayed exactly where they were. Naturally, these changes meant that a new crankshaft had to be used with suitably amended bearing positions.

ENGINE SPECIFICATIONS

There were five sizes of XK engine available, the capacity, bore and stroke of each being shown in this table:

Capacity (cc)	*Nominally*	*Bore (mm)*	*Stroke (mm)*
2483	2.4	83	76.5
2792	2.8	83	86
3442	3.4	83	106
3781	3.8	87	106
4235	4.2	92.07	106

SALOONS

Model	*Engine prefix*	*Capacity (cc)*	*Quoted bhp*
Mk VII	A, AB, BD, D	3442	160
Mk VIIM	D, DN, NA	3442	190
2.4/2.4SE	BB	2483	112
Mk VIII	N, NA	3442	210
3.4 Saloon	KE, KF	3442	210
Mk IX	NL	3781	220
Mk 2 2.4	BG, BH, BJ	2483	120
Mk 2 3.4	KG, KH, KJ	3442	210
Mk 2 3.8	LA, LB, LC, LE	3781	220
Mk X 3.8	2B	3781	265
S-type 3.4	7B	3442	210
S-type 3.8	7B	3781	220
Mk X 4.2	7D	4235	265
420G	7D	4235	265
420 Saloon	7D	4235	245
240 Saloon	7J	2483	133
340 Saloon	7J	4235	210
XJ6 4.2	7L	4235	245
XJ6 2.8	7G	2792	180
XJ6L 4.2	7R	4235	245
XJ6 4.2 S2	8L	4235	245
XJ6L 4.2 S2	8L	4235	245
XJ6C	8L	4235	245
XJ 3.4 S2	7L, 8A	3442	161 (D)
XJ 4.2 S3	8L	4235	170 (D)
XJ 3.4 S3	8A	3442	161 (D)

(D) Horsepower measured to DIN standard rather than gross.

SPORTS CARS

Model	*Engine prefix*	*Capacity (cc)*	*Quoted bhp*
XK 120	W, WF	3442	160
XK 120SE	W, WF	3442	180–190
XK 140	G	3442	190
XK 140SE	G	3442	210
XK 150 3.4	V	3442	190
XK 150 3.4SE	V	3442	210
XK 150 3.4S	VS	3442	250
XK 150 3.8	VA	3781	220
XK 150 3.8S	VAS	3781	265
E-type 3.8	R, RA	3781	265
E-type 4.2	7E	4235	265
E-type S2	7R	4235	265

BUYING ENGINE SPARES

The most obvious place to buy spare parts is from a franchised Jaguar dealer. Though their prices are generally somewhat higher than those of specialist concerns, you can always be sure that you are buying the right quality part for the right car and with a guarantee. Also, you may find that some parts are available at what seems like bargain-basement prices (on occasion, dealers have sales to clear out old and more unusual parts) so it's always worth checking your local dealer.

That said, there are many reputable specialists who can supply a wide variety of maintenance and accessory items either cheaper or in a wider variety. Companies such as SNG Barratt and David Manners have an enviable reputation for providing quality service and body parts at reasonable prices, and they have an excellent computerised parts location and mail order service.

Specialising further still are companies such as the Carburettor Exchange which (no surprise!) concentrates on SU carburettors – a feature of almost all XK engines.

Naturally, as the earlier models get older and less prolific, it's reasonable to expect that the availability of genuine spares will decrease. However, the large number of specialists should ensure that most XK engines can still be kept in good fettle at reasonable cost. Wiser owners will make a point of snapping up more unusual bits and pieces and storing them away – just in case. Getting out to one or more of the dozens of Jaguar events held every year across the country can often be worth it just for the opportunity to pick up such items.

In choosing a supplier, or someone to work on your car, a good reputation is worth a thousand adverts. Word of mouth from a contented customer (or two) is your best guide, as perhaps would be a recommendation from a specialist magazine or one of the Jaguar Clubs or Registers.

BUYING SECOND-HAND

When buying mechanical items, it is often difficult to ascertain whether they are worth the asking price – or, indeed, worth anything at all. Buying, say, a used camshaft (one owner, never raced or rallied), or a cylinder head, demands some real care as it's very hard to know exactly the condition and it's impossible to perform crack tests without special equipment. And, items such as alternators need to be tested under load in order to be sure that they are operating correctly. Remember also that many items, even for the older XK-engined Jaguars, are still available on a new and/or reconditioned exchange basis – for example, just about all parts for SU carburettors and fuel pumps are available either from specialists or direct from the suppliers, Burlen Fuel Systems Limited.

THE KIT FORM ENGINE

Buying an XK engine as a box of bits (the proverbial unfinished project) is a risky business, to say the least. Unless you are really clued up and have lots of time, it will be difficult to ascertain whether all the bits and pieces are actually there. Moreover, you will also need to ascertain whether the components all relate to one engine – with 40 years' worth of XKs around, it could easily be a 'bitza' engine. That 'original' XK 120 engine might just include more than a few bits of XJ6. If you must buy an engine like this, be sure to take an expert along with you when you go to see it.

PART 3

Stripdown and reassembly

Chapter 9

Removing the Engine and Gearbox

LIFTING AND SAFETY

A combined XK engine and gearbox can weigh more than 407kg (8cwt) which means that serious lifting gear is required. Get an engine hoist with sufficient capacity – Clarke make a wide range of different models – and, ideally, have a helper ready to guide the engine out. Drain the fluids before removing the engine/gearbox from the car and never get beneath it once it's on a hoist.

PREPARATION

An hour or two spent in logical preparation could save days of aggravation – and, conceivably, lots of cash.

- Make sure you've enough room to work – allow for the movement of the engine hoist.
- Get the engine bay as clean as possible.
- Have a large selection of containers – boxes, jars, tins, etc – for holding fasteners.
- Keep masking/gaffa tape and a felt-tip pen to hand for marking components.
- Similarly, use 'Tipp-Ex' to mark such things as plug leads or timing marks.

AIR CONDITIONING AND FUEL COOLER SAFETY

If you have a Jaguar fitted with air conditioning and/or a fuel cooler, some extra care is required during engine removal. In some cases it is possible to leave the air-conditioning components connected during the process – check your Haynes Manual for specific details – but if disconnection is required, the system MUST be discharged by a Jaguar dealer or air-conditioning specialist beforehand. It must *not* be attempted on a DIY basis. If your car has a fuel cooler, it cannot be removed before the refrigerant has been professionally discharged.

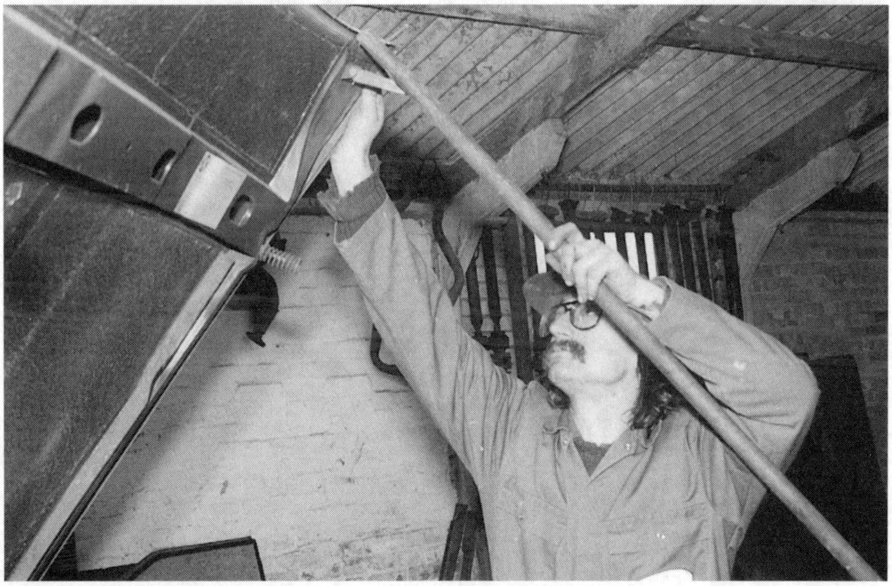

▲ If you are working on your engine while it is still in the vehicle, it is always a good idea to prop the bonnet open where applicable – the springs can get weary on older cars. For engine removal, it's usually advisable, and in some cases necessary, to remove the bonnet altogether. For most XK-powered Jaguars, the bonnet is a particularly large lump of metal and you will need some assistance during removal to ensure you don't damage it. In some cases, such as the E-type, it's technically possible to remove the engine with the bonnet in situ. However, it's much easier without it getting in the way and, given the price of a new one, it's not worth the risk of damaging it.

◀ *Remove the battery earth lead to prevent any accidental short circuits. If the engine is coming out, remove the battery altogether.*

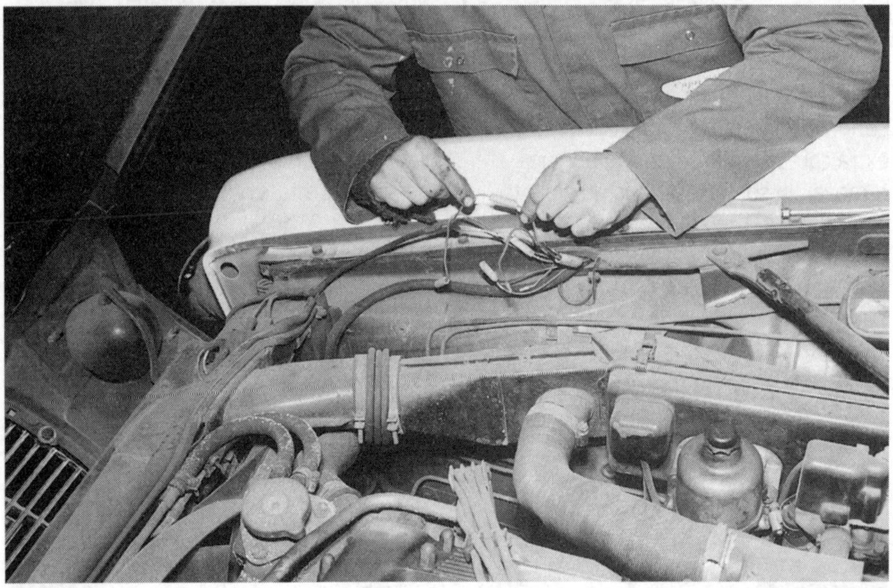

◀ *All wiring from car to engine has to be disconnected and/or removed. It's only when you start to do this that you realise just how much wiring is involved in your car. Use the masking tape and a biro to label the wires as you remove them. Don't make the mistake of thinking you'll remember which is which – there are a thousand operations to go through before you return to this point, and the odds are your memories will be a blur.*

▶ *The later the car, the more electrics will be involved. Check very carefully all the time that you don't miss anything – this electrical multi-plug was tucked away in the dark depths of the engine bay.*

FUEL INJECTION

▶ *Taking out a fuel-injection engine is almost the same as its carburettor'd cousin but with more electrical and air/fuel connections. As previously, it's important to ensure that anything linking the engine to the body of the car is unplugged or disconnected. This is the multi-plug from the ECU (electronic control unit) to the all-important air-flow meter.*

◀ *Though the manifold assembly and fuel rail can remain in place for the engine removal, the air-flow meter has to come off. When removing fuel pipes and connections, it's important to take the usual precautions with regard to fire. Keep fuel off the bodywork and your skin.*

◀ *The carbs can stay in situ, but the fuel line from the tanks has to be disconnected. Fuel can spill out, even after the fuel tap has been closed, so take care. The throttle and choke cables have to be disconnected. (Some models have a start enrichment device.) On an automatic-gearbox car, the kick-down cable has to be removed from the accelerator linkage.*

▲ There are lots of individual wires to be removed as well – the oil sender at the bottom of the engine is easily missed. Some models have a second unit toward the front of the engine. There's also the lead from the starter motor solenoid and the HT and LT cables from the coil and, shown here, an earth cable on the bulkhead.

▶ *Where power steering is fitted, it's often possible to leave the pump attached to the car. In this case the bolt holding the steering pump supporting strap, which is bolted into the water pump at one end and the power steering pump at the other, was removed; as was the fan belt. A long bolt secures the power steering pump. It's possible to ease the pump back out of the way so that the engine can be extracted with the pump still in place. This means that there's no need to remove those tricky-to-refit crimped-on hose connections. If the pump is leaking, an exchange unit is the best option.*

▲ *In theory, the single bolt holding the engine to the mounting bracket should be removed. In truth, it's often hard to get at and virtually impossible to remove without better access. The simpler answer is to remove the bracket itself.*

▲ *Don't forget to release the connectors from the rear of the alternator (dynamo, on earlier models). The exhaust down-pipe connections will almost certainly be seized at the manifold, so it's simpler to release the downpipe clamp where it joins the exhaust system and remove the down-pipe at the engine stripdown stage.*

▲ *Using the Clarke engine hoist, the chain hooks are attached to the lugs on top of the cylinder head and the hoist is given a bit of lift to slightly take the strain off the car whilst the various gearbox connections are released under the car. Obviously, it is important when working under the car to support it with axle stands. In this case, a trolley jack was used to support the gearbox during the process. Easy to forget is the gearshift knob – you can't get the engine/gearbox out if it's left on!*

You'll need to disconnect the propeller shaft at its front flange and tie it securely out of the way, and disconnect the speedometer cable. On manual transmission cars, unbolt the clutch slave cylinder from the clutch bellhousing and tie it out of harm's way. There's no need to disconnect the clutch hydraulic lines. The leads to the overdrive switch (where applicable) should be disconnected.

AUTO TRANSMISSION REMINDERS

On cars with an automatic transmission there are several items to undo or remove, over and above those cars with manual transmission, notably:

- The speed selector cable from the lever on the side of the transmission housing.
- The speed selector cable clamp from the torque converter housing.
- The vacuum pipes and electrical leads.
- Fluid cooler pipes.

▼ *The rear of the engine/gearbox should be lowered until the gearlever clears the console grommet, and then the assembly can be eased forward and hoisted up and out of the engine bay. It has to be a steep angle to clear the front of the car, as you can see here. This is an XJ6 with plenty of clearance at the front. Many Jaguars (the Mk 2, for example) don't have this clearance, which makes engine removal harder work still.*

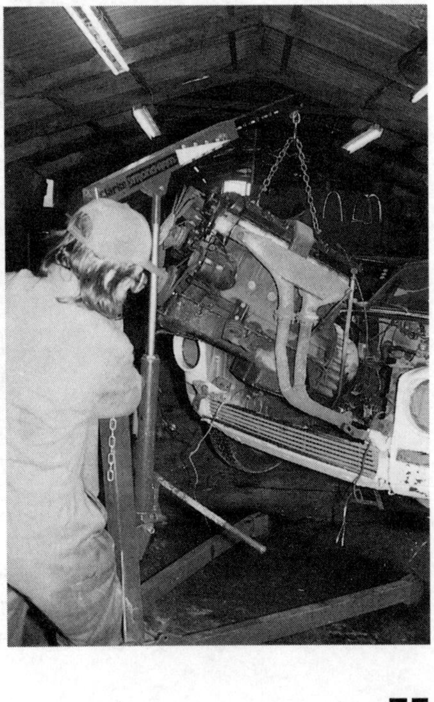

▲ *The weight of the gearbox/engine combination cannot be overemphasised – there's a lot of it! On no account try to manhandle the engine, as back or other injuries are almost a foregone conclusion. Use a quality engine hoist, such as the Clarke model shown in these pictures – if you haven't got one, most hire shops will supply them at a daily or weekly rate.*

REMOVING THE GEARBOX FROM THE ENGINE

▲ *Remove the oil pipes from the gearbox to the engine. Where applicable, remember that the auto 'box dipstick is secured on a stud into the exhaust manifold. The starter motor is held by two long nuts/bolts which pass through the engine/gearbox flanges. There's a large gasket, which should be replaced.*

▲ *On automatic cars the torque converter has to be released. The securing bolts (with two tab washers per bolt) can be seen through the hole left by the starter motor.*

▲ *Now you can remove the gearbox-to-engine mounting bolts, remembering to support the rear of the gearbox to take the strain as more bolts are removed. Note that the top bolt in this photo is missing. This indicates that the 'box has been removed previously with the engine in situ – and, as it is very difficult to reach that bolt on a reassembly, the fitter simply hadn't bothered with it!*

▲ *Seen from the front, the XK 120 unit shows once more the similarities and differences of three decades – plenty of the former and not many of the latter. And, no, on top of the cam covers wasn't the original position for the starter motor!* (Courtesy Rob Beere Racing Services)

◀ *Compare the engine in the previous photos (a Series 2, twin-carb XJ6 unit) with this very early XK 120 engine. Very similar, despite there being over 30 years between them. Note the lack of fasteners around the front of the cam covers – a source of irritation and oil leaks for many owners until the factory added three fasteners to keep them tight. Note also the oil filter assembly – many different types have been used over the years.* (Courtesy Rob Beere Racing Services)

Chapter 10

Engine Stripdown

This section – dealing with the stripdown, investigation, repair, replacement and, finally, the rebuild of the engine – was photographed at the premises of VSE Engineering Ltd. Tim Camp's highly professional operation deals solely with the restoration of XK engines and sales of XK spares. Tim has raced and rallied many Jaguars, with tuned XKs of various hues providing the power.

Between Tim and his crew, one very tired 4.2-litre engine (in this case, from an XJ6) was transformed from an oil and travel-stained caterpillar into a resplendent black and silver butterfly. Whilst there are differences between this and other XK units, they are generally in the details, and the basic procedures shown here apply to all models. In the interests of simplicity, detailed specifications have been restricted to the most popular XK applications. My grateful thanks to Tim and the crew for their words of wisdom, passed on to you now.

▶ *The VSE team. Left to right they are: Malcolm Standen, Russell Kemp, Doug Whittingham, Neil Arnold and Tim Camp.*

▶ *Russ started by taking off the exhaust manifold heat shield. Where applicable, the 'steady' for the auto transmission filler/dipstick is fitted to the lower rear stud. Most of the fasteners will be rusted, and easing agent will be needed. Take your time, because stripping threads on the studs will only add to your grief. If you're having trouble, ease the nut out a fraction of a turn then back again, to work some of the accumulated rust free of the threads. Tap the manifold gently with a rubber mallet to break it free. The two 3-cylinder manifold gaskets should always be replaced. VSE also replaces the nuts and studs with stainless equivalents, thereby pre-empting similar problems at a later date.*

BITS AND PIECES

When rebuilding your own engine, it is vital to store all fasteners and components so you know *where* they are AND *what* they are. Make a note of sequences and the positioning of washers, etc – it may seem obvious now, but in three months time when you're rebuilding the engine, you may not be quite so sure. Taking photographs is a good way to aid your memory. Even better, the family video recorder, when tripod mounted, is ideal as a visual and aural aid, as you can record verbal notes to explain any particularly intricate sections.

▲ *Russ removed the engine mounting left in place earlier – there is a flat and a spring washer underneath the nut – and then the three bolts holding the engine mounting bracket to the engine.*

▲ *He removed one of the bolts from the alternator adjusting rod. (Earlier cars used dynamos fitted in the same location.) This steel bolt goes into the aluminium engine timing cover at the front of the engine and, with the inevitable electrolytic reaction, it often seizes. He slackened the adjusting rod bolt where it fits on to the alternator, and removed the single lower alternator mounting bolt. It's still quite a snug friction fit, so Russ could easily steady it with one hand while he removed the drive belt, before easing the alternator away. The drive belt is always replaced and the alternator is performance-tested. Drive belts, and even alternators, are generally available, even in high street stores like Halfords.*

Four bolts hold the alternator mounting bracket. You can't remove bracket and alternator as a single unit, as it hides one of the bolts. The top left-hand bolt requires the use of a ring spanner. Two bolts are short and two are slightly longer because the inner section of the bracket is double-skinned.

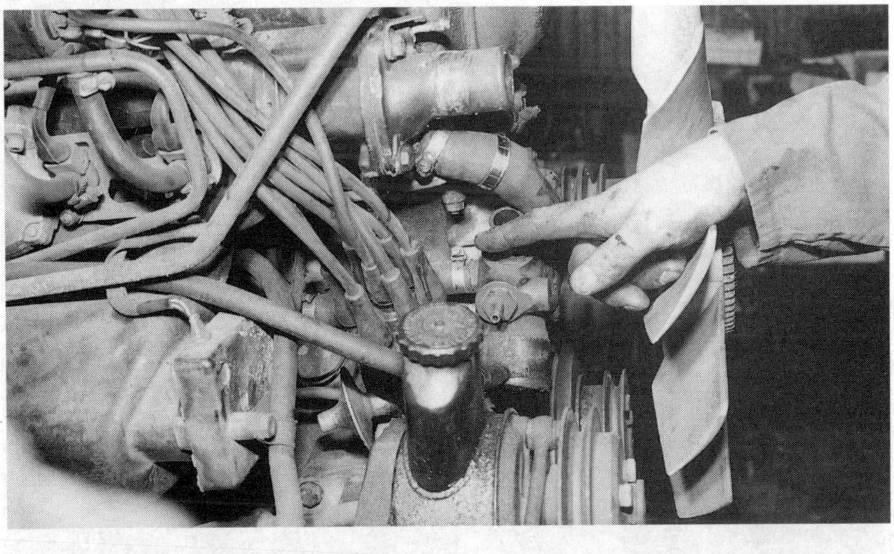

▲ *To remove the water pump, Russ first took off the large hose which leads to the bottom of the radiator and the three others, not forgetting this small one at the back of the pump that connects to a metal pipe which goes to the heater. Check the condition of all the hoses, bending them back and examining for hairline cracks. Five bolts secure the water pump. There are two studs at the top and one at the bottom. When you've removed them all you can …*

◀ *... pull the water pump off the block, complete with the fan assembly. As you can see, there is plenty of rusty residue in both the pump itself and the aluminium timing cover. This often causes corrosion in the latter, and in extreme cases it allows the oil and water to mix, thus displaying all the classic symptoms of a blowing head gasket.*

▲ *To remove the front pulley, you need to lock-up the flywheel. VSE have a special tool for doing just that and it fits as shown here. With the nut removed, the damper should follow, but it often sticks. A conventional 3-leg puller would pull the damper off the rubber inner so ...*

▲ *... VSE also developed a special puller which does the job quickly and pulls the whole assembly off as a piece.*

▲ The pump is held by a single long bolt and nut. The bolt holding the stay to the water pump had already been removed (in the engine removal section). Russ supported the weight of the pump as he withdrew the bolt – note the locating plate on the end. There is a steel insert in the rearmost hole of the pump mounting bracket and a steel collar which goes on the outside to aid the location of the bolt.

HIGH TENSION TALK

If you're not replacing them, mark your plug leads with 'Tipp-Ex' or similar – BEFORE removal! For reference, No. 6 cylinder is at the front and No. 1 at the back.

If you are re-using the distributor, again use 'Tipp-Ex' to mark its exact position and save a lot of time setting-up. A single bolt secures the distributor. Unplug the wiring from the front of the unit and remove the inlet manifold vacuum pipe, and then waggle the distributor out of the block. Take care not to use excessive force which could damage the components.

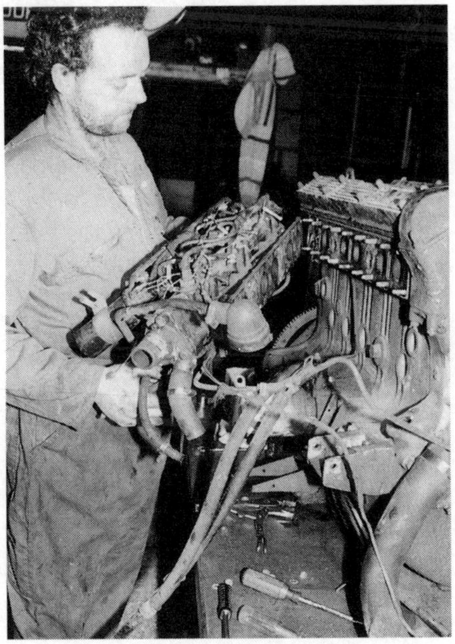

On this XJ6 engine (and others) the cylinder head breather pipe passes from the front of the engine into a rubber pipe on the inlet manifold. Removing it means that the plug leads can be taken off all at once, rather than separately.

Next to remove are the coil and parts of the wiring loom situated close to the manifold. This engine is fuel-injected, so there's a lot more 'gubbins' to remove before you can actually get at the inlet manifold. Ensure all electrical connections have been disconnected before you pull the manifold off.

Remove the two brackets securing the steel section of pipe from the water pump to the heater. Then disconnect the wire going to the oil pressure gauge sender and likewise the other going to another sender at the front of the engine which cuts fuel supply if pressure drops too low. The latter can prevent the engine from starting and can be the cause of mysterious no-go problems.

◀ The manifold, secured by 18 nuts and spring washers, will probably require a tap from a rubber mallet. Take care not to damage that soft aluminium cylinder head.

◀ Carburettors are a little easier to remove, although there's still plenty to do. The carbs shown in this section are SU HS8 – note that with these (and the later HIF7 carbs), the auxiliary carburettor had been replaced by an AED (automatic enrichment device), which is the black plastic box sitting between the carbs. The sequence here shows the removal of these carburettors from an engine that has already been removed from the car, with basic instructions assuming it is in the vehicle – adapt where required. Remove the air cleaner, including back plate and filter element, close the tap on the fuel filter and disconnect the fuel line banjo unions for the carburettor float chambers. Disconnect the throttle cables and the leads from the choke solenoid. On cars with auto transmission, remove the spring clip which secures the kick-down link at the back of the rear carburettor. Disconnect the throttle return springs. Shown here is the removal of the hot air delivery pipe.

◀ You've also got to remove the crankcase breather pipes and …

▲ … the distributor vacuum advance pipe.

▲ Remove the four nuts which secure each carburettor to the inlet manifold – the top inner nuts are a little tricky to get at. Note that there isn't room to get the ring of this combination spanner on the nuts at all four corners.

▲ The carbs will almost slide straight off the manifold mounting studs, but it may be necessary to ease the carburettors away from the manifold before the last two nuts can be reached to release them from the last few threads. If the carbs stick on the manifold gasket, a sharp tap with the flat of the hand on the smooth dashpots should do the trick. The two carbs and the AED, complete with linkages, can be lifted clear all-of-a-piece and taken to the bench for examination, and stripdown if required. Always renew the gaskets or O-rings located on the flanges, whichever type of sealing device is used.

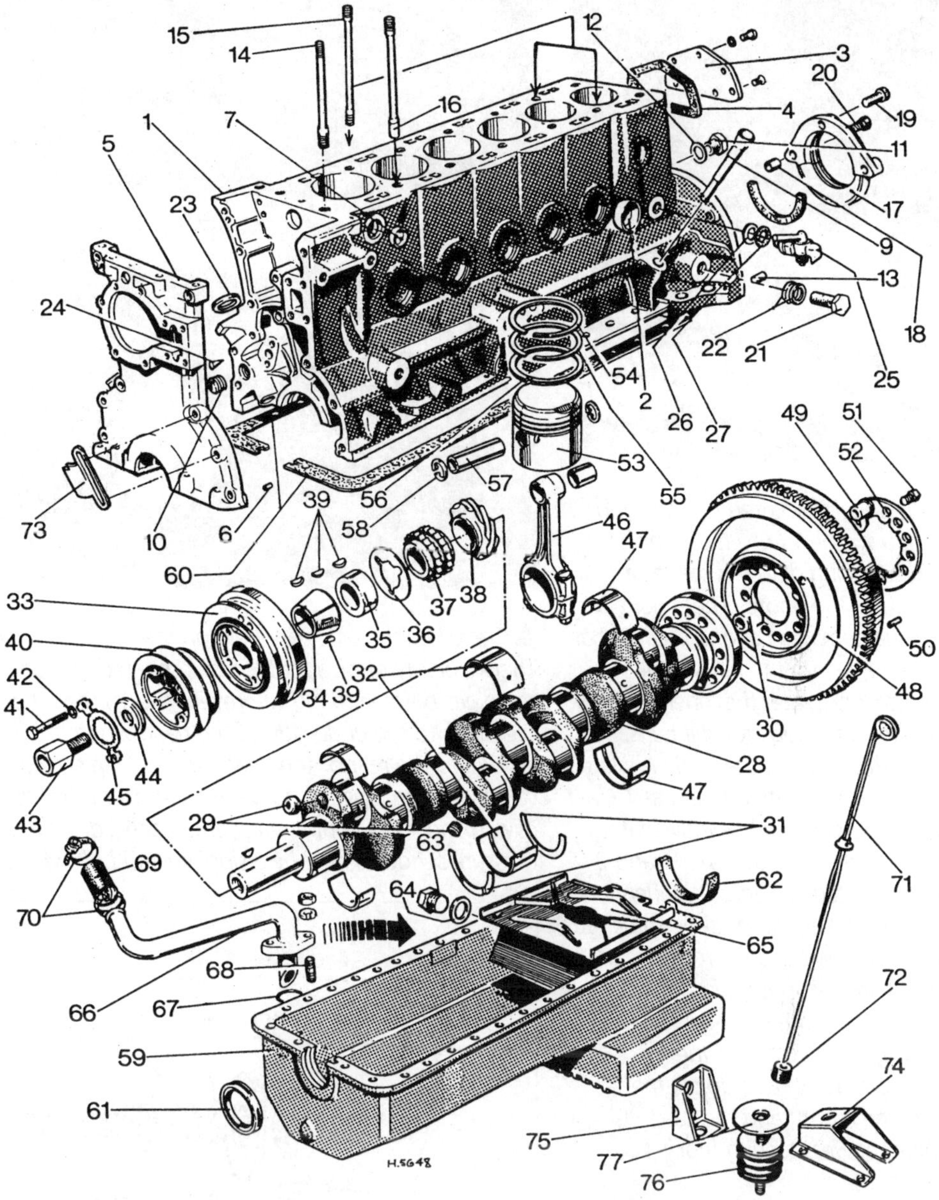

1. Cylinder block
2. Core plugs
3. Blanking plate
4. Gasket
5. Timing cover
6. Dowel
7. Core plug
9. Dipstick tube guide
10. Plug
11. Hexagon plug
12. Copper washer
13. Dowel
14. Stud
15. Stud
16. Dowel Stud
17. Rear oil seal cover
18. Dowel
19. Socket headed screw
20. Socket headed screw
21. Banjo bolt
22. Copper washer
23. Sealing ring
24. Filter gauze
25. Drain tap
26. Copper washer
27. Fibre washer
28. Crankshaft
29. Plug
30. Bush (Clutch pilot)
31. Thrust washer
32. Main bearings
33. Damper
34. Cone
35. Distance piece
36. Oil thrower
37. Crankshaft sprocket
38. Oil pump drive gear
39. Woodruff key
40. Pulley
41. Bolt
42. Lock washer
43. Bolt
44. Washer
45. Lock plate
46. Connecting rod
47. Big-end shell bearing
48. Flywheel
49. Dowel
50. Dowel
51. Setscrew
52. Lock plate
53. Piston
54. Top compression ring
55. Lower compression ring
56. Oil control ring
57. Gudgeon pin
58. Circlip
59. Sump
60. Gasket
61. Crankshaft front oil seal
62. Crankshaft rear oil seal (part)
63. Drain plug
64. Copper washer
65. Baffle assembly
66. Oil pump pipe
67. O-ring
68. Stud
69. Hose
70. Hose clip
71. Dipstick
72. Ferrule
73. Ignition timing point
74. Engine mounting bracket
75. Engine mounting bracket
76. Flexible mounting pad
77. Washer

▲ *As you can see here, there's plenty of work still to do in order to get the XK unit down to its component parts!*

▲ *The flywheel is secured by bolts and locking tab washers (which cannot be re-used). The locking device is still in situ at the top left of the photo. These bolts are high-torque items and a long breaker bar is required. Beneath the bolts is the single piece lock plate, the inner sections of which form the tabs for the bolts. To the left in this photo is the packing plate – note the larger locating holes, at 12 and 6 o'clock. The flywheel/driveplate locating dowels fitted to Series 1 and 2 cars are not fitted to Series 3 models. Russ used a socket, flat washer, threaded bolt and nut to wind it out from the threaded thimbles that go into the back of the crank, and the flywheel came gently away. Russ removed the bellhousing bracket (two bolts), the other engine mount and the cast aluminium power steering bracket.*

▲ *The oil-cooler pipes are held on the block by a plate secured by a single nut. There's plenty of oil about, so be prepared for some spillage. Always plug the ends of the pipes before storing to prevent the ingress of dirt. The pipe leads up the side of the block to the underside of the head where it is clipped into place. It continues around to the rear of the head where a T-piece and extensions carry oil to each camshaft. The pipe has to be released from the clip and then the bolts removed. The olive gland has a copper washer at each side of it, which should be replaced. The oil pressure sender is held by a single gland bolt which again has two copper washers.*

▶ *The oil filter was unscrewed and discarded. This diagram shows the various parts involved in a typical filter assembly*

1. *Base*
2. *Sealing ring*
3. *Element*
4. *Canister*
5. *Sealing washer*
6. *Centre bolt*
7. *Spring*
8. *Relief valve*
9. *Seal*
10. *Felt washer*
11. *Washer*
12. *Plate*
13. *Spring clip*

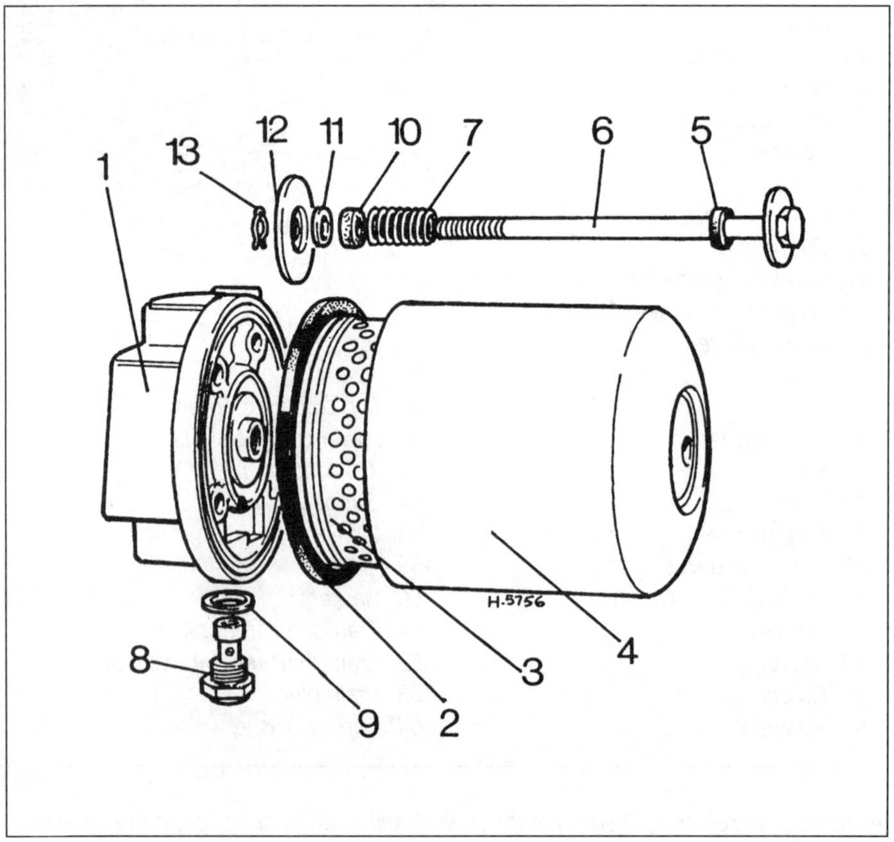

▲ *The oil filter drain-back pipe, which carries the oil back down to the sump, is secured by two bolts. Russ removed them, and then the four bolts holding the filter housing to the block, so that the whole assembly could be removed as a piece. There's a rubber sealing washer on the pipe where it goes into the sump.*

▲ *The cylinder head breather is secured by four domed nuts with washers underneath. You should find two gaskets sandwiching a gauze filter. It's fairly common to find that both gaskets have been replaced by a silicone 'bodge'.*

TIMELY REMINDERS

Before removing the timing pointer from the cylinder block, Russ cleaned up the area and marked its position by gently making a mark across the edge of the pointer and the timing cover. He then removed the two bolts and the pointer (these bolts are two of many holding the timing chest cover in place).

▲ *The rocker covers were then removed. These were secured by a differing number of fasteners over the years, so it's important to make sure you've got them all. The domed nuts should have copper washers underneath – if they're not there on stripdown, make sure they are when you rebuild. The covers may stick a little; bearing in mind the earlier comments, a gentle tap with a rubber mallet should ease them free. There's a notch at the front of each cam, seen in this photo at 12 o'clock. The cams should be aligned so that the lobes point outward like 'rabbit's ears'. When the head is removed it can be placed on a flat surface with the minimum of protrusion from the valves. Each cam is held to its sprocket by four bolts, with a locking tab for each pair of bolts (earlier models had locking wire). Only two bolts can be removed at once. Turning the crank gives access to the remaining two bolts.*

▶ *On this engine it's easy to forget the six nuts holding the front of the cylinder head in place. Two have already been removed (they secured the fuel pipe and alternator wiring loom) but there were still two at either side, together with suitable washers.*

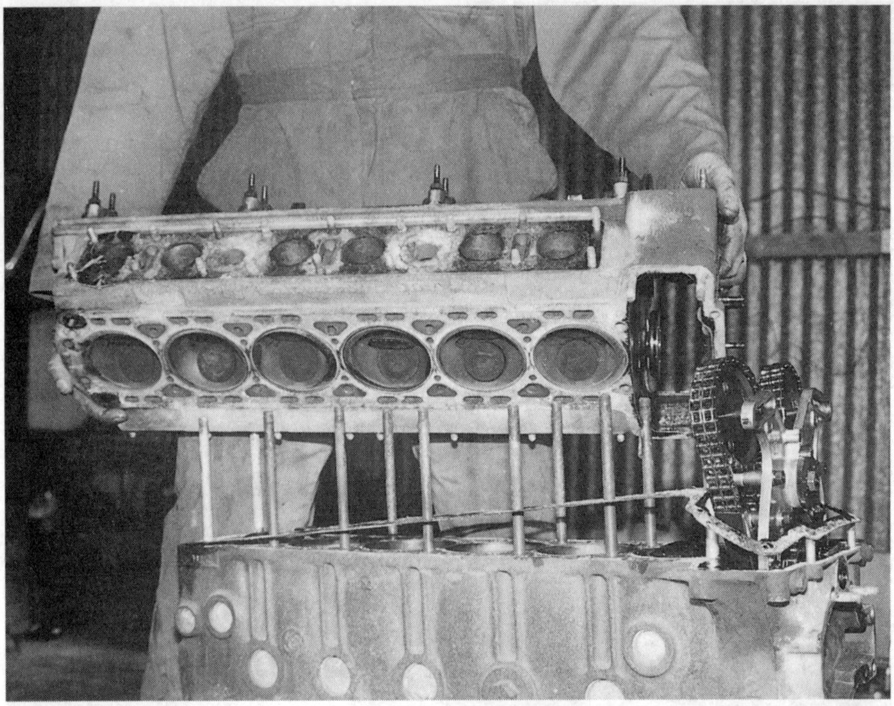

▲ There are 14 domed nuts to be removed and, to prevent head warpage, it's important that they are tackled in the right order – a reversal of the order for tightening them (see diagram elsewhere in this chapter). There should be one washer underneath each nut, but studs stretch over time and it's common to find more than that on some studs. They also rust because they sit in the water jacket of the cylinder block, so wind them out slowly and carefully! When they've all been removed, the head can be lifted free – a tap from a rubber mallet will probably help. If it really sticks hard, then prolonged (and ever more forceful) hammering will do nothing but expensive damage. It is sometimes possible to use a bottle jack, suitably positioned, to push the head off.

▲ At this point, Malcolm Standen took over. He took off the sump, which is held by bolts all the way around, except at the front which is four studs, nuts and washers. At either side of the rear seal housing, the bolts are much longer than the others. Note also the rear main oil seal Torx screws. He used the engine hoist to raise the block high enough to remove the sump, and then laid the engine on its RH side to avoid damaging the dipstick tube. Don't damage the pick-up pipes in the sump, and it mustn't rest on cam sprockets at the top. Malcolm then removed the various brackets which secure the oil pick-up pipes and laid them on the bench.

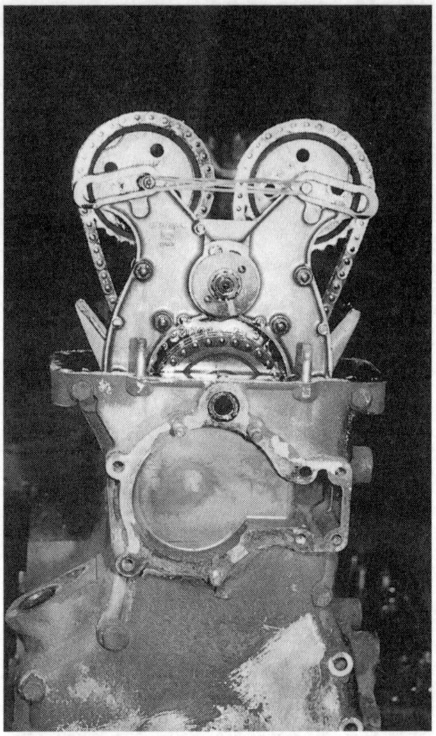

▲ Russ removed the cylinder head gasket and used an elastic band to hold the cam sprocket assembly in one piece.

▲ The all-important oil pump is secured by three bolts – two are shouldered and one is part threaded. VSE always replace the pump – it's a small item that can wreck an engine. With the oil pump off, Malcolm then removed the pump drive assembly. This bronze bush comes off first, and the rest can follow once the locking tab on the drive pinion has been tapped back. This is the drive assembly in its entirety. Make sure you don't damage the (relatively) soft bronze bush during the removal process.

This diagram shows clearly the component parts of the oil pump, pick-up pipes and drive assembly.

1. *Pump body*
2. *Rotor assembly*
3. *Cover*
4. *Setscrew*
5. *Setscrew*
6. *Spring washer*
7. *O-Ring*
8. *O-ring*
9. *Distributor driveshaft*
10. *Bush*
11. *Washer*
12. *Gear driven from crankshaft*
13. *Woodruff key*
14. *Nut*
15. *Special washer*
16. *Coupling sleeve*
17. *Dowel bolt*
18. *Dowel bolt*
19. *Tab washer*
20. *Oil delivery pipe*
21. *Gasket*
22. *Tab washer*
23. *Strut*
24. *Clip*
25. *Oil pick-up pipe*
26. *Clip*
27. *Strut*
28. *Strut*
29. *Strut*
30. *Sealing plate*
31. *Spring*
32. *Split pin*

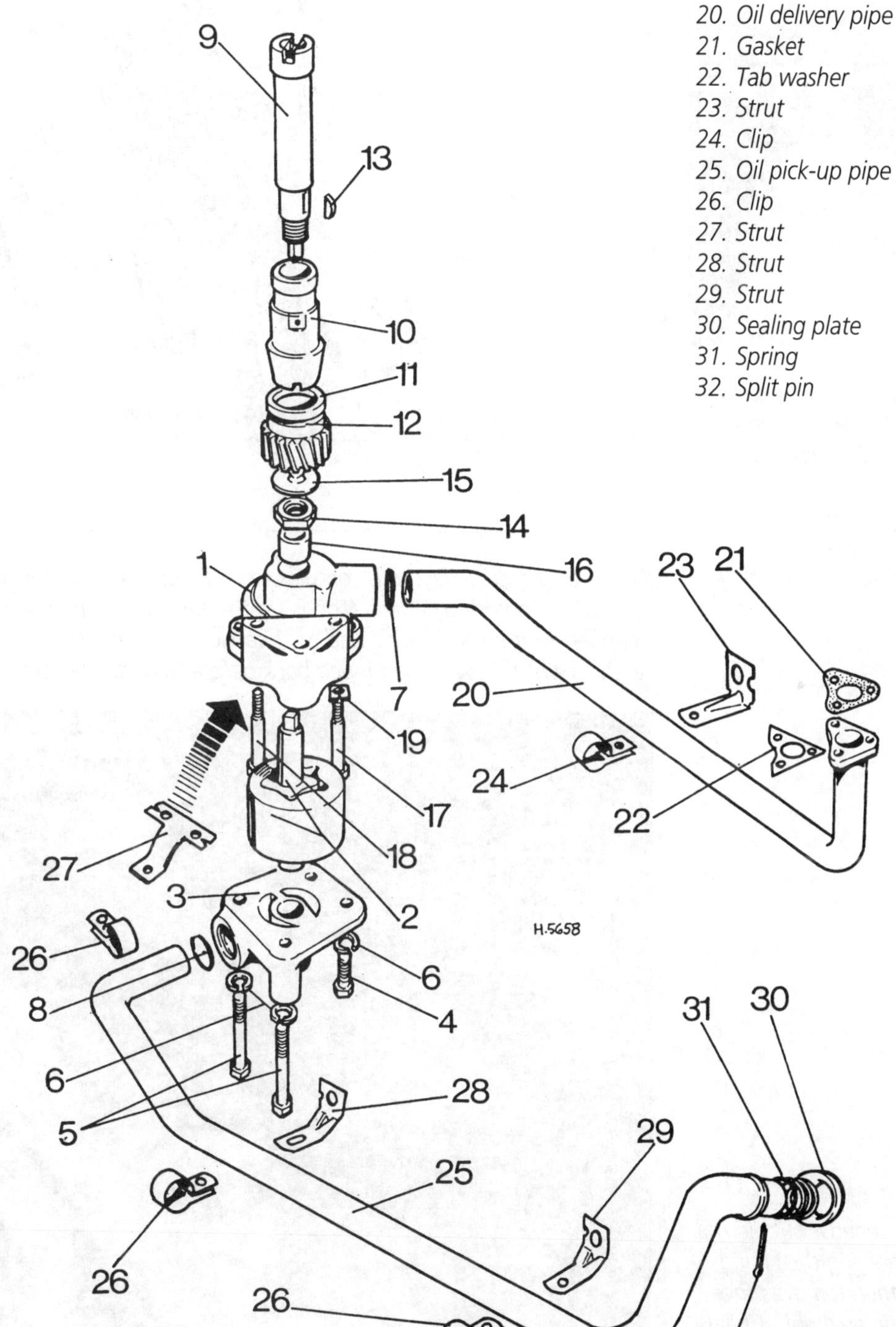

▲ *Malcolm tapped off the front damper cone revealing the Woodruff key in the end of the crankshaft. This has to be tapped free and stored safely. The timing chest cover is held by three ½in bolts and five 9/16in bolts. Malcolm had to tap the casing with a rubber mallet to break it free of the gasket.*

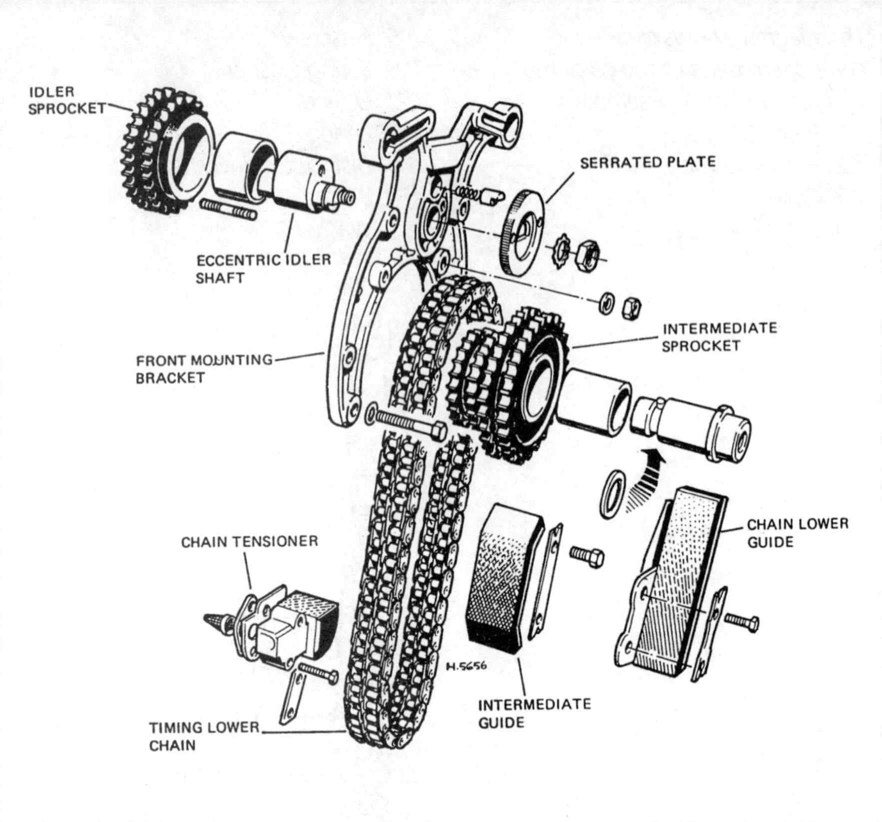

▲ *The timing chain layout is complex, so it's important to have a good grasp of what goes where before you start. (This is for the XJ6.) There are two chain tensioners – a 'manual' intermediate guide and a hydraulic tensioner, and both feature tabbed washers which have to be tapped back before they can be removed.*

▶ *Malcolm then loosened – but did not remove – the bolts holding the timing cover top plate. He then removed the complete timing chain assembly. Chains are always replaced at VSE, though it's rare to find a sprocket with any serious wear problems.*

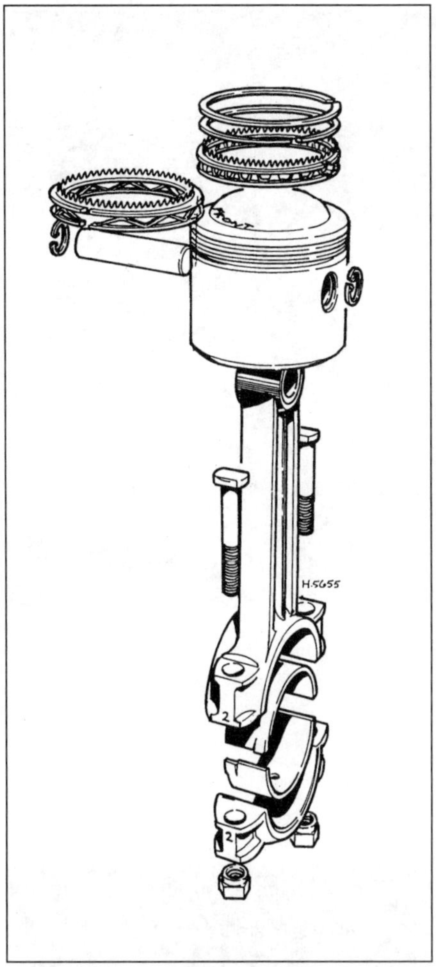

▲ Note that the con rods and caps are numbered, with the highest figure (6) being at the front of the engine. The numbers should always align as shown here.

▲ Malcolm turned the crank so that he could easily access the two nuts on No. 1 con rod. Note that the nuts have double the number of 'flats' (something referred to by many tool suppliers as 'bi-hex'), to ensure that a socket must be used for removal. He removed the cap and checked out the bearing for signs of wear, usually this is a bronzing effect. It's usually wise to replace all such bearings as the cost is minimal. If you are going to re-use them, before storing them away wipe off the oil and stick on a piece of masking tape, marked with their location. Keep each piston and con-rod assembly together – if nothing else, it helps identify problem areas.

▶ The main bearing caps are held by 14 nuts and locking tab washers (always discarded at VSE for safety's sake). There should be a number stamped into the caps or alongside them in the crankcase. If there isn't, use a dot punch to make an identifying mark so they can be matched at rebuild time. The centre cap is the only one to feature a thrust washer at each side. With all the bearing caps removed, Malcolm lifted out the heavy crank ready for examination.

▶ *A small drift was used to tap out the engine oil dipstick tube, and a large socket was required to undo the rear gallery oil plug. There should also be a copper washer which should be discarded. Seven bolts hold the rear water chamber plate, and when it's removed there's almost certain to be some more sludgy rust.*

▲ *The upper rear oil seal housing is held by three Torx head bolts. Note the hollow locating dowels which fit into the cover at either side. Malcolm then turned the engine upright and removed the water drain bolt and copper washer from the rear left of the engine.*

▲ *Removing the studs is a nerve-wracking business, as they often rust badly and shear. Malcolm is using the S-P stud remover – essential for this job. The 14 studs are not all the same length, so take careful note of which goes where.*

▶ *In post-1970 XK engines, the 'long stud' blocks are so designed as to keep the studs in contact with the coolant all the time. The result of keeping steel in constant contact with water is a foregone conclusion, and it's by no means unusual to find studs as badly corroded as the one shown here (an unaffected stud is shown alongside it for comparison).*

▶ *VSE always recommend that the core plugs are removed after the studs, because the plugs could foul on some of the studs as they are tipped out. There are five plugs on the RH side and six on the left. Tap them at the edges to make them tip out, rather than in their centres. Uncared for engines can silt almost all the way up, particularly from the back of the engine. You should use only top-quality anti-freeze, with anti-corrosion inhibitors, all year round to prevent the build-up of nasty sludge like this. The amount of sludgy, rusty (so-called) coolant we've seen so far in this engine bodes very badly indeed.*

There are four oil gallery plugs and copper washers on the RH side of the engine and an oil gallery plug alongside the distributor drive at the front. This requires a large Allen key.

CLEANING

Having reduced the XK engine to its component parts, it's almost time to start work in earnest – but not with all that oil, rusty water and gunge still in place. As we've already seen, VSE automatically binned many parts during the stripdown process, notably where their replacements were relatively cheap and/or where excess wear had occurred. Many other items, however, were to be re-used, but certainly not in the condition they came off the car. Getting items clean is an essential part of any engine rebuild, as it enables the builder to spot any cracks or other problems which may not be immediately obvious – and, of course, who wants a *dirty* rebuilt engine?

For the most part, a proprietary cleaner and/or degreasant will do the job well enough. That and a little hard work with a stiff brush. If you work regularly with dirty car components, a Clarke washer is a good investment as it combines a purpose-made sink with a motor which pumps the cleaning agent directly on to the workpiece. Some pieces, however, are too dirty for standard cleaning methods to work, which is why VSE has an acid bath which cleans up items such as timing sprockets so that they look like new.

▶ *Here are several parts of the timing chain assembly and the con rods going into the VSE acid bath.*

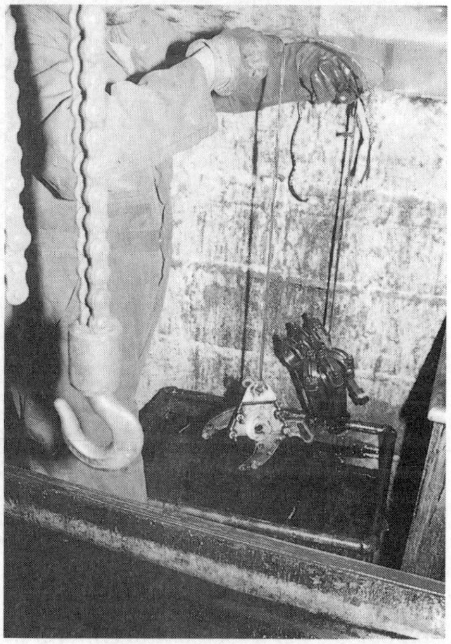

◀ *Smaller items were put into a cage or can for safety – no one was volunteering to grope around in the acid for lost bits of engine!*

▲ *A special aluminium cleaner is used on such things as sumps, with decades of ingrained oil and road dirt to remove, it's amazing to see …*

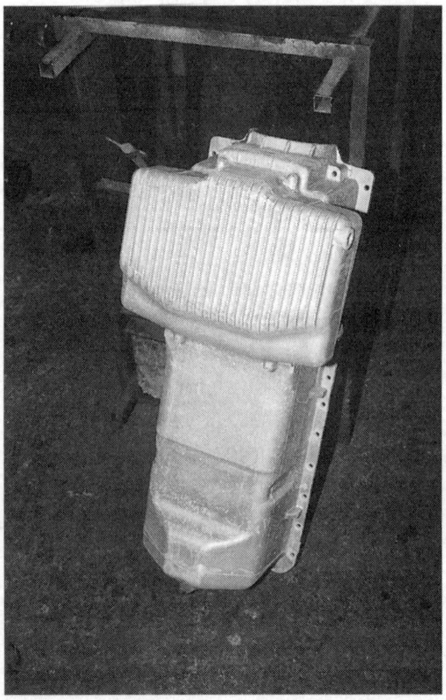

▲ *… the finished result. No, it's not a new one, honestly!*

Chapter 11

Cylinder Heads

There's no doubt that much of the XK engine's success can be attributed to its advanced head design. Being aluminium (DTD 424 alloy to be precise) it is, of course, lighter than a cast-iron unit by around 60 per cent. This lack of weight made it easier to produce, not only in terms of machining, but also from the simple point of view that it was much easier to manhandle around the factory. As already mentioned, much of the head's efficiency came courtesy of Harry Weslake's development work. Hemispherical combustion chambers and advanced valve port design were his stock in trade, which he plied to perfection on the XK.

The original 3.4-litre XK engines (as used in the XK 120 and the Mk VII saloon) featured 1¾in inlet and 1⁷⁄₁₆in diameter exhaust valves. As valve sizes increased along with power outputs, these became known retrospectively as the 'small valve' or 'A-type' cylinder heads. Other features were the spring blade type of timing chain tensioner and the (relatively) low camshaft profile.

When the C-type racing car arrived on the scene, much of its extra performance was derived from modifications made to the cylinder head (notably bigger valves and valve angles) which, understandably, became known in its own right as the C-type head. This was modified slightly for production engines and made available as an optional extra on the XK 120, XK 140 and even on the Mk VII saloon. When first offered on the XK 120 cars, no external markings were used, but when fitted to the XK 140 and beyond, a cast letter 'C' could be seen in the sparking plug well.

The B-type cylinder head followed and was basically a slightly more civilised version of the C-type, confusingly leaving us an initial sequence of A, C and B. At this point, the heads from the XK 120, XK 140 and Mk VII could be changed between engines as there was little real difference and all used the same inlet manifold.

STRAIGHT TORQUEING

The B-type mixed and matched, using the narrow inlet ports of the XK 140 engine, but with the larger inlet valves of the C-type head. There was no improvement in either torque or power but, importantly, the torque peaked 1,000rpm lower down the rev range, thus obviously improving standing start acceleration and mid-range, in-gear performance.

The D-type head was yet another derivation of the C-type, using bigger valves and being fitted to the D-type racing cars (and also the XK SS).

Initially, the XK 150 was fitted with a B-type head, but with the advent of the 'S' models (in both 3.4- and later, 3.8-litre format) the straight port head was introduced. This featured a triple SU carburettor configuration and obviously meant that using this type of head with a twin-SU set-up meant swapping the inlet manifold accordingly. The XK 150S, 3.8-litre engine found its way more or less unchanged into the mighty E-type and, introduced at the same time, the heavyweight Mk X saloon. However, though it could be assumed that these two applications would be identical, they are not, largely because of the lack of room under the E-type bonnet – the inlet manifold exited the head at an angle in order that it didn't foul the bonnet. In addition, the E-type differed in that it was fitted with a manual choke arrangement.

The 420 was the first straight port head to be fitted with two (rather than three) SU carburettors, which increases the scope for easy tuning of less powerful engines.

From the time of its introduction in 1964 (again in the E-type and Mk X) the 4.2-litre engine was only available in straight port guise. It's interesting to note a basic difference between the 3.8-litre and 4.2-litre straight port heads; the inlet manifold of the former was a three-part affair with an external water passage. On the latter, the manifold was a single casting that included, as part of its design, the water passage, which removed the need for the bungs/core plugs required on the 3.8-litre heads.

At the end of the 1960s, Jaguar considered the future of the XK engine carefully and they introduced a number of changes which affected the last few Series 2 E-types (the Series 3 used the 12-cylinder unit, remember), the latter 420/420G and, of course, the XJ6. Longer head studs were fitted in the redesigned block, which itself was extended to include a couple of extra water cooling holes.

HEAD GAMES

In principle, any head from any XK engine should fit any other XK engine. But, not surprisingly, given some 40 years of change and development, it's not quite as easy as that, even though the stud pattern remained the same throughout that period.

CYLINDER HEAD DETAILS

This table shows the details of cylinder heads fitted as standard to XK-engined cars.

Car	*Head ref.*	*Exhaust valve size (inches)*	*Camshaft lift (inches)*	*Compression ratio*
XK 120	A (C type optional)	$1\frac{7}{16}$	$\frac{5}{16}$	8.0:1
XK 120SE	A (C type optional)	$1\frac{7}{16}$	$\frac{3}{8}$	8.0:1
Mk VII	A	$1\frac{7}{16}$	$\frac{5}{16}$	8.0:1
Mk VIIM	A	$1\frac{7}{16}$	$\frac{3}{8}$	8.0:1
XK 140	A	$1\frac{7}{16}$	$\frac{5}{16}$	8.0:1
XK 140SE	A (C type optional)	$1\frac{7}{16}$ (A) or $1\frac{5}{8}$ (C)	$\frac{3}{8}$	8.0:1
2.4	A	$1\frac{7}{16}$	$\frac{5}{16}$	8.0:1
2.4SE	A	$1\frac{7}{16}$	$\frac{5}{16}$	8.0:1
3.4	B	$1\frac{5}{8}$	$\frac{3}{8}$	8.0:1
Mk VIII	B	$1\frac{5}{8}$	$\frac{3}{8}$	8.0:1
XK 150 3.4	B	$1\frac{5}{8}$	$\frac{3}{8}$	8.0:1
XK 150 3.4 SE	B	$1\frac{5}{8}$	$\frac{3}{8}$	8.0:1
XK 150 3.4S	Straight port	$1\frac{5}{8}$	$\frac{3}{8}$	8.0:1
XK 150 3.8	B	$1\frac{5}{8}$	$\frac{3}{8}$	8.0:1
XK 150 3.8S	Straight port	$1\frac{5}{8}$	$\frac{3}{8}$	8.0:1
Mk IX	B	$1\frac{5}{8}$	$\frac{3}{8}$	8.0:1
Mk 2 2.4	B	$1\frac{5}{8}$	$\frac{3}{8}$	8.0:1
Mk 2 3.4	B	$1\frac{5}{8}$	$\frac{3}{8}$	8.0:1
Mk 2 3.8	B	$1\frac{5}{8}$	$\frac{3}{8}$	8.0:1
240	Straight port	$1\frac{5}{8}$	$\frac{3}{8}$	8.0:1
340	Straight port	$1\frac{5}{8}$	$\frac{3}{8}$	8.0:1
E-type 3.8	Straight port	$1\frac{5}{8}$	$\frac{3}{8}$	9.0:1
E-type 4.2*	Straight port	$1\frac{5}{8}$	$\frac{3}{8}$	9.0:1
E-type 4.2 SII*	Straight port	$1\frac{5}{8}$	$\frac{3}{8}$	9.0:1
Mk X 3.8	Straight port	$1\frac{5}{8}$	$\frac{3}{8}$	8.0:1
Mk X 4.2*	Straight port	$1\frac{5}{8}$	$\frac{3}{8}$	8.0:1
420G (*)	Straight port	$1\frac{5}{8}$	$\frac{3}{8}$	8.0:1
S-type 3.4	B	$1\frac{5}{8}$	$\frac{3}{8}$	8.0:1
S-type 3.8	B	$1\frac{5}{8}$	$\frac{3}{8}$	8.0:1
420 (*)	Straight port	$1\frac{5}{8}$	$\frac{3}{8}$	8.0:1
XJ6 4.2 (*L)	Straight port	$1\frac{5}{8}$	$\frac{3}{8}$	9.0:1
XJ6 2.8 E	Straight port	$1\frac{5}{8}$	$\frac{3}{8}$	9.0:1
XJ6 4.2 S2 (*L)	Straight port	$1\frac{5}{8}$	$\frac{3}{8}$	9.0:1
XJ6 3.4 S2 (*)	Straight port	$1\frac{5}{8}$	$\frac{3}{8}$	8.8:1
XJ6 4.2 S3 (*L)	Straight port	$1\frac{5}{8}$	$\frac{3}{8}$	8.7:1
XJ6 3.4 S3 (*L)	Straight port	$1\frac{5}{8}$	$\frac{3}{8}$	8.8:1

NOTES:
(*) Cylinder head with staggered bores.
(*L) Cylinder head with staggered bores and long studs.
(E) Cylinder head with equidistant bores.

INLET VALVE SIZES

The inlet valves on all XK engines were $1\frac{3}{4}$in diameter with the exception of the XJ6 4.2-litre S3 cars, where the diameter was increased to $1\frac{7}{8}$in.

CYLINDER HEAD COLOURS

Over the years, Jaguar developed a system of colour-coding the cylinder heads. Sympathetic restorers, such as VSE, who supplied the following list, will always stick to it.

Car	Colour
XK 120	Unpainted
XK 140 standard	Unpainted
XK 140 'C' head	Red
XK 150 3.4	Duck egg green
XK 150 3.8	Blue
XK 150S 3.4/3.8 (triple carburettor)	Gold
Mk VII/VIIM	Unpainted
Mk VIII	Duck egg green
Mk IX	Blue
Mk X	Gold
E-type 3.8	Gold
E-type 4.2 SI (polished cam covers)	Gold
E-type 4.2 SI/SII (black/silver ribbed cam covers)	Unpainted
420G (polished cam covers)	Gold
420G (black/silver ribbed cam covers)	Unpainted
420 (black/silver ribbed cam covers)	Unpainted
Mk 1 & 2, 2.4/3.4, 3.4S	Duck egg green
Mk 2 3.8/3.8 S-type	Blue
XJ6	Silver

CAUTION: When buying a replacement head, do not rely solely on its colour (or lack of it) as a means of identification – not everyone is scrupulous, and a wily seller could easily paint a head a suitable colour to suit your needs, and his pocket!

It's interesting that all 4.2-litre engines were fitted with straight-port cylinder heads and that with the advent of the 4.2-litre engine, Jaguar didn't bother to change the basic head casting; the result being that they never *quite* align properly.

In theory, because the heads from all models are so similar, it is possible to swap and change quite easily. To some extent this is true. For example the 240 and 340 models used the same straight port head as the 420, but with respectively HS6 and HD6 SU 1¾in carburettors. As such, an easy way to obtain some extra urge is to do the straight swap for a 420 head, complete with its 2in SU HD8 carburettors. Equally, this head can be fitted to Mk 2 engines. Clearly, the power-hungry will already be thinking in terms of fitting a triple-SU carburettor head, but there just isn't room in the engine bay to fit them all in.

Heads from the late 1960s are generally known as carburettor heads (they were designed at the time of the introduction of the XJ6), although they were also fitted to the last of the Series 2 in E-types, 420 and 420G saloons. Because they included extra water holes to suit the uprated block, switching and changing these heads becomes problematical, though not impossible – the holes (either in the block or head, depending on the swap taking place) can be filled, but more than a little thought should be given to this. Better still, take advice from one of our two engine experts.

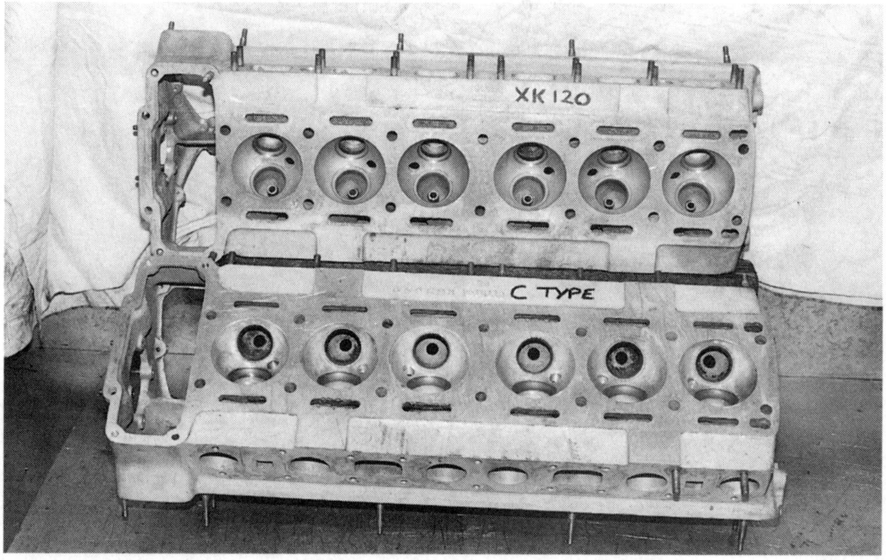

▲ *At top is an XK 120 head with its C-type, big valve derivative below.* (Courtesy VSE Engineering)

▲ *A newly skimmed XK 120 complete with all components necessary for a rebuild.*

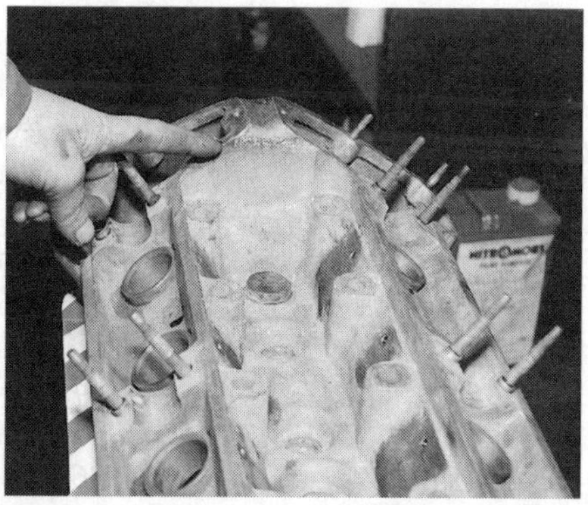

▲ *Head identification. If the cylinder head has a polished front – as here – then the engine number will be stamped into the rear of the front section (and it is repeated on the cylinder block). The ribbed cam cover, cast-front XJ6 engines have a reference number stamped into the back of the cam valley.*

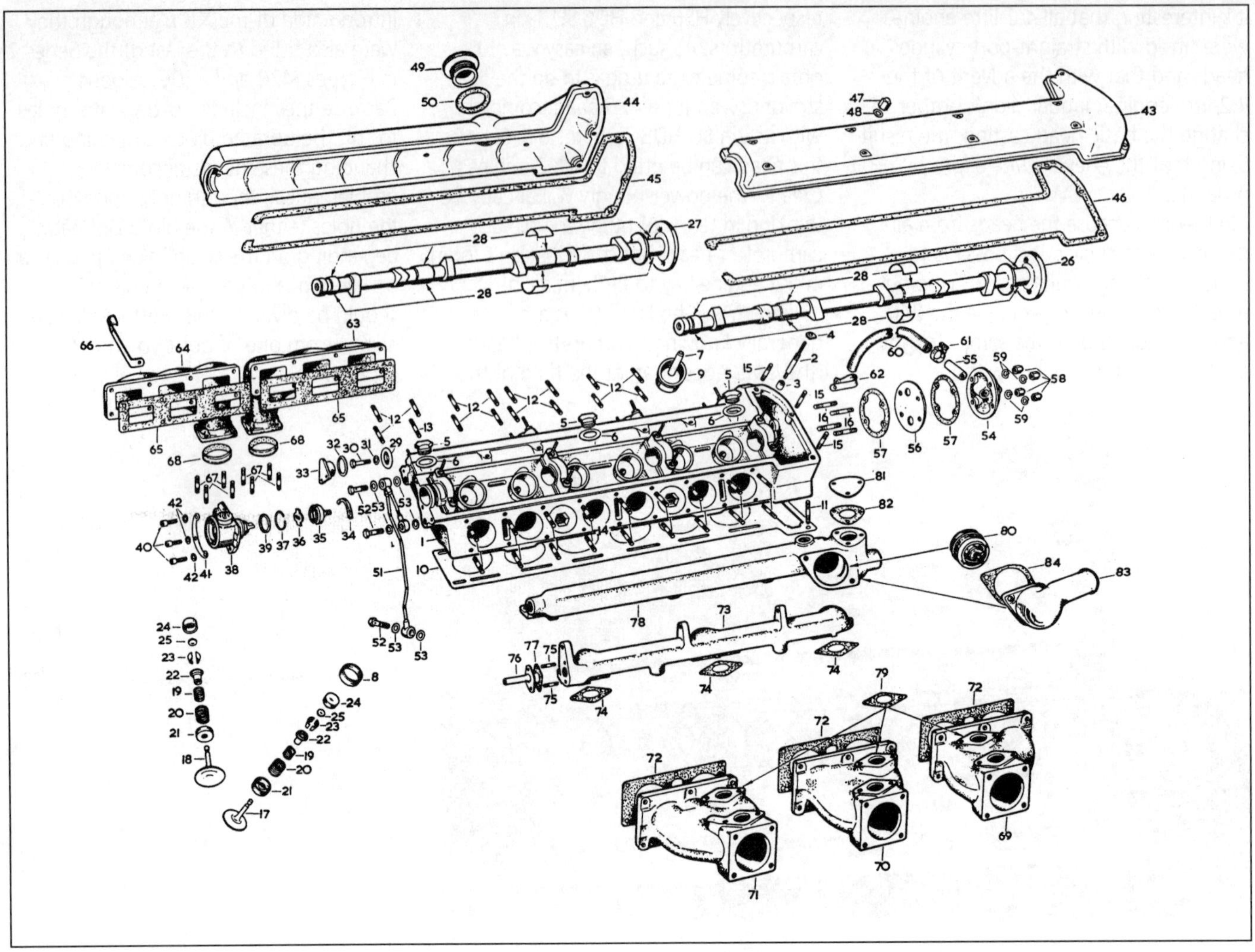

Whichever model XK you have, the head is a complex assembly, as can be seen here in this exploded view of an E-type head.

1. Cylinder head
2. Stud
3. Ring dowel
4. D washer
5. Plug
6. Copper washer
7. Valve guide
8. Valve insert
9. Tappet guide
10. Gasket
11–16. Stud
17. Inlet valve
18. Exhaust valve
19. Valve spring
20. Valve spring
21. Seat
22. Collar
23. Cotter
24. Tappet
25. Adjusting pad
26. Inlet camshaft
27. Exhaust camshaft
28. Bearing
29. Oil thrower
30. Set screw
31. Copper washer
32. Sealing ring
33. Sealing plug
34. Seal
35. Adaptor
36. Driving dog
37. Circlip
38. Generator
39. Sealing ring
40. Screw
41. Plate washer
42. Lock washer
43. Inlet camshaft cover
44. Exhaust camshaft cover
45. Gasket
46. Gasket
47. Dome Nut
48. Copper washer
49. Filler cap
50. Fibre washer
51. Oil pipe
52. Banjo bolt
53. Copper washer
54. Breather housing
55. Pipe
56. Baffle
57. Gasket
58. Dome nut
59. Spring washer
60. Flexible pipe
61. Clip
62. Clip
63. Exhaust manifold
64. Exhaust manifold
65. Gasket
66. Clip
67. Stud
68. Sealing ring
69. Inlet manifold
70. Inlet manifold
71. Inlet manifold
72. Gasket
73. Air balance pipe
74. Gasket
75. Stud
76. Adaptor
77. Gasket
78. Water pipe
79. Gasket
80. Thermostat
81. Plate
82. Gasket
83. Elbow
84. Gasket

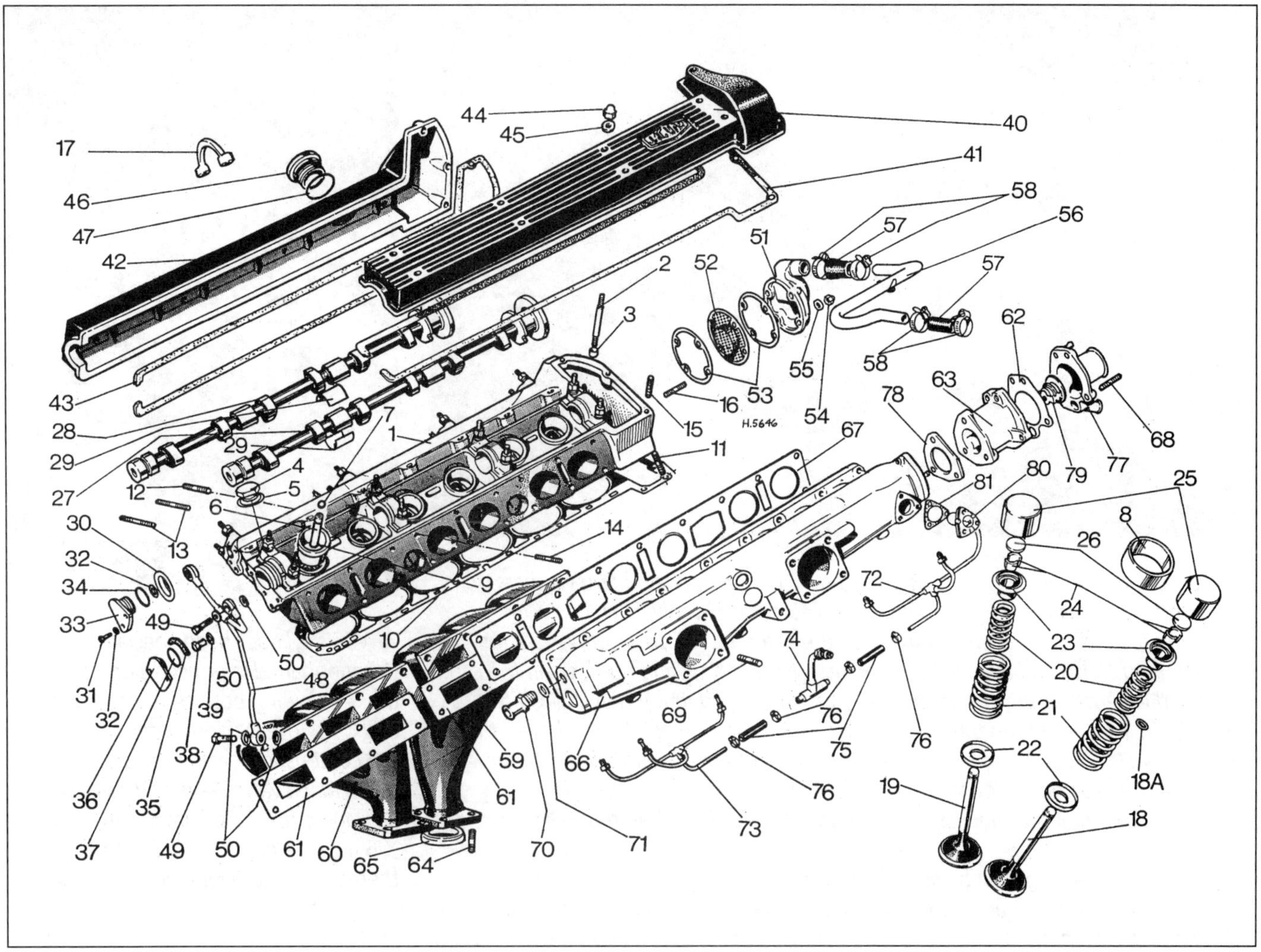

Compare that last diagram with the final incarnation of the XK engine as fitted to the XJ6. Though there are detail changes, its instantly recognisable as an XK.

1. Cylinder head
2. Stud
3. Dowel
4. Core plug
5. Washer
6. Valve guide
7. Valve guide circlip
8. Inlet valve insert
9. Tappet guide
10. Cylinder head gasket
11–16. Stud
17. Lifting lug
18. Inlet valve
18a. Seal
19. Exhaust valve
20. Valve spring (inner)
21. Valve spring (outer)
22. Valve spring seat
23. Valve spring collar
24. Split cotters
25. Tappet
26. Valve adjusting shim
27. Inlet camshaft
28. Exhaust camshaft
29. Camshaft bearing
30. Oil thrower
31. Setscrew
32. Washer
33. Plug
34. O-ring
35. Seal
36. Plug
37. O-Ring
38. Setscrew
39. Washer
40. Camshaft cover
41. Gasket
42. Gasket cover
43. Gasket
44. Dome nut
45. Washer
46. Oil filler cap
47. O-ring
48. Oil pipe
49. Banjo bolt
50. Washer
51. Breather hosing cover
52. Flame trap
53. Gasket
54. Dome nuts
55. Spring washers
56. Breather pipe
57. Hose
58. Clip
59. Exhaust manifold
60. Exhaust manifold
61. Gasket
62. Gasket
63. Thermostat housing
64. Stud
65. Sealing ring
66. Inlet manifold
67. Gasket
68. Stud
69. Stud
70. Adaptor
71. Washer
72–74. Manifold starting pipe
75. Neoprene tube
76. Clip
77. Water outlet pipe
78. Gasket
79. Thermostat
80. Automatic choke thermostat
81. Gasket

▲ There are four main caps holding each cam, with two nuts and two washers on each. Undo the nuts vertically so that the cam comes up evenly. Each cap is numbered to correspond with a number on the head, from 1 to 8, where 1 is the LH front (exhaust front) and 5 is the inlet side rear. Note the casting numbers in the cam.

Inlet and exhaust cams are different. They're not marked, but when in position correctly the lobes will both turn outward to form 'rabbit's ears'. Crossed over, the cam lobes will both point inward. The ends of the cams wear, particularly if the engine has been standing for some time and the cams have been touching the followers. It's common to find cams wearing on the sides just before the base circle starts.

The eight cam bearings can be tapped out using a suitable drift. VSE always replace them. Russ removed the cam followers with a magnet. If they stick, tap the top of the follower with a hammer. If worn, they must be skimmed and re-hardened. Unless you are doing a complete head overhaul job, keep the tappets and tappet shims in pairs. A spring compressor and small screwdriver can be used to ease out the two collets per valve. If they're not too badly worn, they can be re-used.

▲ VSE's cylinder head specialist is Doug Whittingham. He noticed signs of a gasket blowing between Nos. 4 and 5 cylinders. The leaking water gets into the combustion chambers, up behind the exhaust valve seats and causes damage. This is a common fault in the XJ6 4.2-litre engine, not least because there is minimal metal between the cylinder bores.

◀ Left to right, each valve assembly comprises: valve, rubber stem seal (inlet valves only), spring seat, two springs (inner and outer) valve top and collets (at front). Early valve tops are deeper, spring seats are deeper and collets are longer. Early XK engines do not have valve seals and this can lead to smoking on start-up (and an unwarranted reputation for being smokers). The answer is to use a later spring seat, top collet and valve seal – inlet only. VSE always fit new valves, springs and seals. Russ dipped the head into an acid bath to make baked-on oil and carbon easier to remove – the spring seats, collets and valve tops were also put in for a good cleaning.

◀ *When things get really bad, you end up with a head that looks like this. The water corrosion damage is obvious just about everywhere, and there's a burnt valve (far left) just to add to this customer's joy.* (Courtesy VSE Engineering)

▲ *Another typical XK head problem is that of hairline cracks down in the cam valley. There's one here which has been highlighted by a felt tip pen. As you can see, it's working its way towards a core plug. Professional repair is called for in this case. It can be welded, but this calls for massive amounts of skill and no small amount of luck. Stitching – drilling a series of small holes and inserting screws into it – is easier and safer than welding.*

▲ *Two inlet valve seats (Nos. 4 and 5 cylinders) needed attention on our head. The area around the seat was cleaned up, and Doug drilled out the two valve guides to ensure that the dummy valve used by VSE would slide easily through it. This is used to knock out the seat in the following procedure. Because it has to be hammered down the guide, the end of the valve gets burred over and won't come out of a standard diameter guide. Old inlet valves are dropped into the other guides to prevent welding spatter getting in there. Doug then applied a neat layer of weld to the seats.*

As the welds cooled, they shrunk the seats and he was able to tap them out. In cases like this, if the recess has not been too badly damaged, it can be machined to take an oversized seat. If it is, then weld will have to be applied to the head to return enough metal back to facilitate machining to the correct size.

The edges of the two valve seat recesses were chamfered slightly and smoothed with emery cloth so that the seat would slot into place more easily. To tap the seats into place, the head has to be heated to around 150°C (302°F) – no problem here, as it had to go into the VSE oven anyway to remove the valve guides (see later). Doug applied a little oil to the edge of the new seat and then tapped it into place using another special VSE tool. A feeler gauge was used to make sure the seats were fully down and equal all the way around. This work has to be done quickly while the head is warm and the seat is cold.

◀ No. 5 seat is in position, but No. 4 required an oversize seat (shown here) and had to be machined to suit. In cases like this, the outside diameter (OD) of the seat is slightly bigger but the inside diameter (ID) is standard. Exhaust and inlet valve seats are both set at a 45 degree angle.

▼ One method used to prevent a recurrence of this problem is to bolt a bracket in between the cylinders. However, this only holds the guide right at its edge, and in extreme circumstances it can move around.

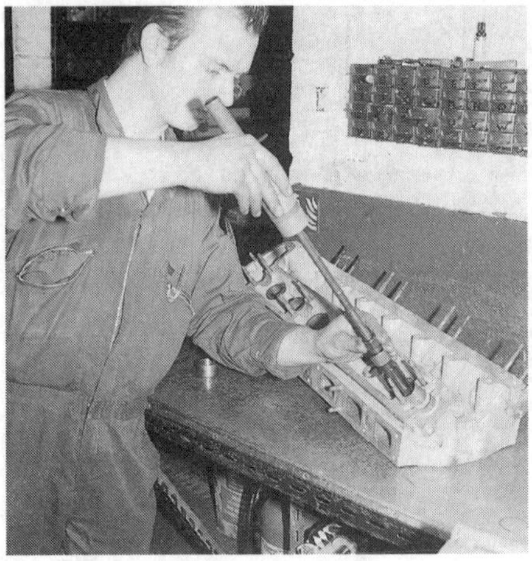

▲ As mentioned, the bucket guides can lift up out of their locations as a result of water damage from a leaking head gasket. Doug removed the damaged bucket guide by using a slide hammer with a special extractor head. Another special tool was used to tap home a replacement bucket guide.

▶ A better method is to drill a hole through from the inside of the valley into the bucket so that a grub screw can be inserted to secure it firmly. A taper tap was used first, and when the thread was complete the hole was filled with Loctite 648 to make a good seal and secure the screw. Doug screwed the grub screw into place and filled the hole with plastic metal. VSE's massive experience of developing racing XK engines has shown that this is the best way to keep the bucket guides where they should be.

▶ *This is a Rob Beere Racing Services 3.8-litre gasflowed racing head with bronze seats, valve guides and tappet guides, as bronze gives better wear and heat dispersion qualities. It's been designed to run with 1.5in tappets and a full race, high lift cam.*

▶ *Inlet guides have two circlips and exhaust guides have one. The circlips are tapped on to the guides and act as locators as the guides go into the head. On the inlet guides, one groove is for the circlip, the other is for the oil seal (no oil seal on the exhaust side, of course). An oversize guide should have markings below the circlip grooves. The guides on the right of this photo are special silicon bronze items designed for serious performance engines.*

To remove the guides, the head has to be heated to 150°C (302°F). Doug makes it a point to always check the diameters of the guides with a micrometer, just to make sure they are not oversize already. Although they should be marked, not all are.

If an inlet guide is loose as it is inserted, the head must be reamed out for the fitting of an oversize guide. When the guides have all been fitted, the head must be allowed to cool slowly and evenly. Once cool, any oversized guides must be reamed out to the correct ID and the others reamed out very gently to ensure that there are no burrs from the tapping-in process. (Courtesy VSE Engineering)

▶ *It's important for the angles and the valves to be correct. The two new valve seats had to be cut into the correct angle, using this Mira machine. New guides had been fitted and so the Mira was used to confirm that the guides, valves and seats were at the right angle, and also to clean up the valve faces.*

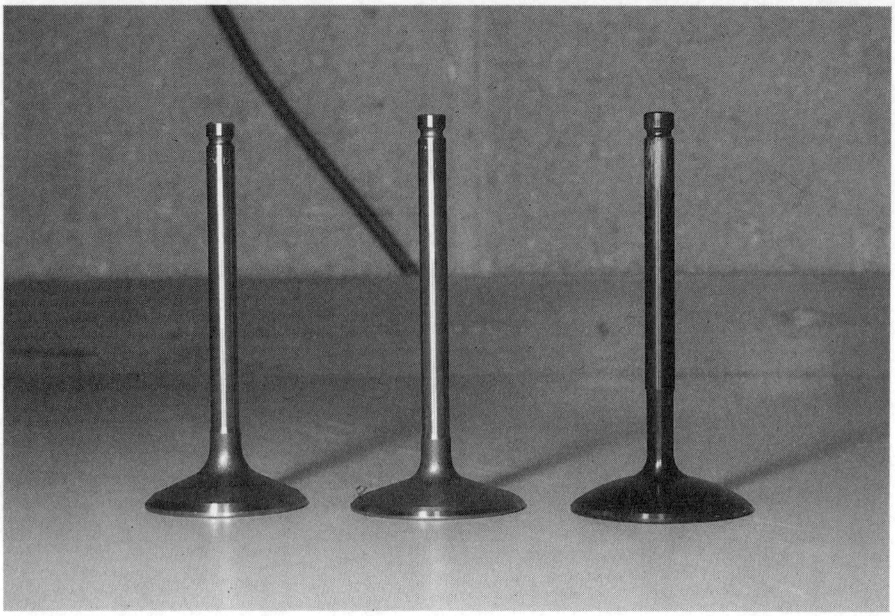

▲ *All valves are not the same. Rob Beere Racing uses valves to suit the application. On the left is a standard 1¾in valve, in the middle is the larger 1⅞in valve (as fitted to the Series 3 XJ6 cars) and at far right is a 1.9in, 214N grade, tuftrided stainless steel valve specially made for the company to fit to racing engines. When the head and/or cam has been uprated in any way, it is recommended that uprated valve springs (both shorter and stronger) are fitted.*

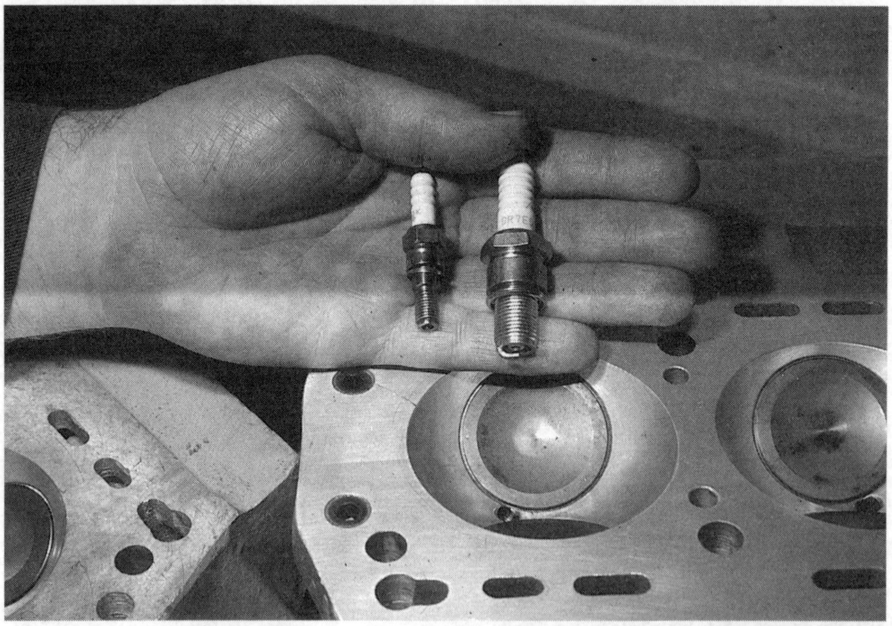

▲ *As a company very much involved in producing successful racing engines, RBRS came up with an unusual innovation, shown here. The original spark plug holes were aluminium welded and finished off, and new ones were made and positioned almost to the centre of the chamber. The holes were small because 8mm motorcycle surface discharge plugs were used. The increase in efficiency was massive and it was so much more extra power that it was banned from its class! The head shown here is coloured green because it was designed for use with an army tank. (All tank heads are green.)*

▲ *'Our' head was then taken to the machine shop where the combustion chambers were thoroughly cleaned up, and then the head was skimmed. A steel straight edge and feeler gauges were used to ensure that it was absolutely flat. Having removed the head, it is false economy just to assume that it is all-square – the odds are that it isn't. Clearly, putting a warped head back in place (even if the warping is only minuscule) is going to lead to gasket sealing problems, which in turn will lead to the kind of water damage problems we've seen earlier in this section.*

▲ Doug fitted the valves into the lapping tool and painted a layer of fine grinding paste around the very edges before inserting the whole lot into the head in order to match each valve to its seat. Always keep the paste off the valve stems, or they will soon wear away. Here's the finished article, the last perfectly matched valve and stem.

▲ These are the components required to build up the valve assemblies.

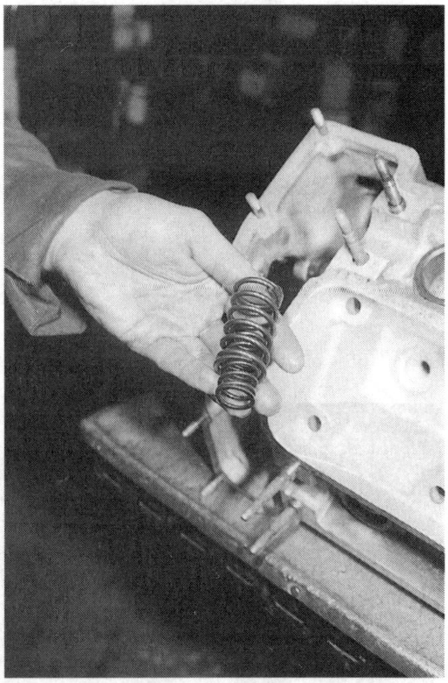

▲ All valves should be oiled before being slid into their respective holes. Because this is a fuel injection head, the springs are heavy duty. Note that there are two springs – the outer one is a standard spring and the inner one is a damper spring. The valve top was fitted and then the valve spring compressor was applied. In order to make fitting the small collets easier, Doug put a dab of grease on the end of a small screwdriver, and stuck them to it. Simple, but effective, and it also gives a little start-up lubrication. Always ensure that the groove in the collet aligns with the groove in the valve stem.

Once Doug had fitted all the valve assemblies, he upended the head and inserted the four studs at the front of the block. This is so that, when the head is tipped the other way up for cam fitting, the rotation of the cams causing the valves to move in and out won't foul them on the bench top.

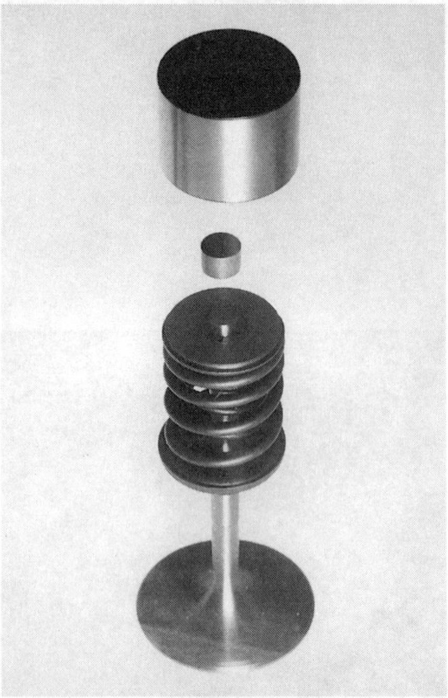

◀ *This photo shows clearly how the valve and its associated components should be assembled.* (Courtesy Rob Beere Racing Services)

FITTING THE CAM ASSEMBLY

▶ *VSE always fit new Glacier cam shells whenever a cylinder head is rebuilt. Re-using the originals, no matter how good they may appear, is false economy, having put in this amount of work. The shells are put into both the head and the caps 'dry'. They are numbered 1–8.*

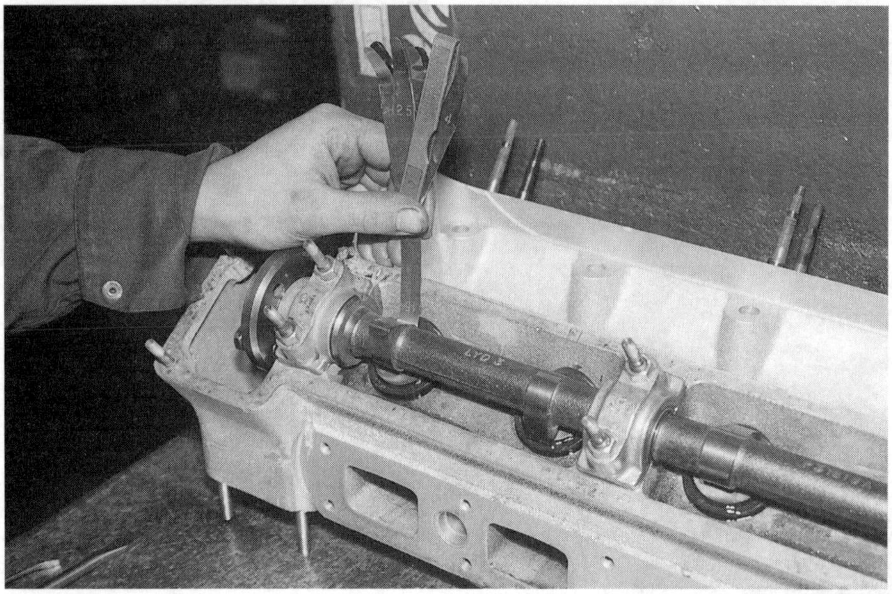

▲ *The shims were then put on to the tops of the valves. The right shim value is vital. Doug usually starts with 85/95 thou and adjusts from there. The shims must be clinically clean, as dirt and swarf on the surface could throw the figures way out. The valve assemblies are always filled with engine oil before the new buckets are installed, using the magnet which was used to remove them.*

Doug fitted two bolts into the end of the cam in order to achieve some leverage once it's bolted into place. Doug applied a drop of oil to the tops of the buckets and then laid the cam in place with the front lobes pointing outward, like 'rabbit's ears'.

IMPORTANT NOTE: When setting up one cam, make sure the other is out altogether to prevent the valves colliding. Doug oiled the shells in the caps before putting them in position and tightening them up evenly. He then checked, double-checked and re-checked the valve clearances on each valve until they were absolutely spot-on.

▲ *Having got the shimming correct, Doug added the special washers under the nuts and tightened the cam down. Take care here, as they are alloy and can easily be damaged – they should only be tightened to 9lb/ft, which is little more than hand tight.*

▲ *Different cams are required for different applications, remembering that the wilder the cam (ie the more lift it has) the more power will be available – all things being equal – but also the more demanding it will be to drive. As a rule, general road use, especially on the UK's crowded roads, will be made particularly unpleasant if too high a lift cam is fitted.*

Seen here at Rob Beere's workshop is a standard cam on the left, with a slightly reprofiled cam in the centre (note the decrease in cam diameter which is compensated for by increasing the shims). On the right is a race cam, made especially for RBRS. This is steel, while the other two are chill cast, offering an incredible 50 per cent weight-saving, despite the increased diameter. Another interesting facet is that Jaguar usually make the hole down the centre of the cam by drilling twice – once from each end. Life being what it is, they don't always meet up exactly. However, the race cam seen here is drilled in one pass for RBRS by a company more used to drilling gun barrels, and so the accuracy is 100 per cent.

▲ *Shown here on the left is a sintered steel Jaguar tappet (follower) and alongside is a 'top hat' steel one used by Rob Beere in full race engines (typically with the race cam shown earlier). The race shim fits over the end of the valve stem and is 'sucked' into place by the oil. The object being that when the engine is revving to its limit, there is no danger of the shim moving around and getting wedged in the follower, as can happen with standard items.*

▲ *At this point the cam could be set-up accurately using this special gauge. This can be purchased outright or hired from one of the three Jaguar clubs (if you're a member!). VSE always fit new manifold studs, 18 on the inlet side and 16 on the exhaust side. Note that the inlet studs are shorter.*

▲ *Polishing the chambers is an essential part of creating better engine efficiency and, ultimately, more power, torque and even fuel economy. It's important to get the air and fuel swilling round evenly to create a cleaner 'burn' at the spark. (Ironically, the inlet ports are not made too smooth, as it has been found that the fuel tends to run down the surface in droplets rather than atomising into the fine mist essential for good combustion.)*

▲ *Having burned the fuel, getting rid of the exhaust gasses is another aspect of tuning, to which end the exhaust ports are also polished. This way the gasflow from the head has a smooth path to run and thus is quicker and more efficient.*

▲ *Russ shot-blasted the old covers to clean off all the original paint and to make a good key for the heat-resistant, competition black Spectra spray paint. Then he ran a 3M Scotchbrite CSD-S plastic paint stripping wheel across the tops of the covers to remove the paint on the ribs, producing a final effect that looks straight off the production line at Browns Lane.*

▲ *VSE can produce a similar factory-look for earlier cam covers, too, including those that are coloured (see separate chart). Note the obvious differences between these two covers: on the left is the very early 8-stud type which tended to leak oil from the front as they couldn't be fastened down tight enough. The 11-stud cover on the right was introduced in 1952 and has more holes drilled in it to allow three more nuts to hold it in place.* (Courtesy VSE Engineering)

UNLEADED FUEL

According to Tim Camp, all XK engines can run on unleaded fuel as long as the heads are equipped with hardened valve seats. The Series 3 XJ6 had these as standard, but VSE can easily fit them to XK engines not so fitted. The ignition timing has to be altered slightly and, in some cases, the carburation, too. In general, Tim finds that better performance results from using the current Super unleaded fuel, although this depends very much on the individual engine.

Chapter 12

The Cylinder Block and Crankcase

▲ *Comparison of blocks can tell you a lot. Check out the bore position on this 4.2-litre block and …* (Courtesy VSE Engineering)

▲ *… this 3.8-litre unit. As you can see, this one is effectively 'two threes', rather than the equally spaced bores of the larger engine.* (Courtesy Rob Beere Racing Services)

▲ *Having got the block down to its basics, a sharp blade should be used to remove the excess gasket material from the bottom of the block. At VSE they soak them overnight in a large 'bath' of paraffin-based solvent, and then they spray-clean with water heated up to 60°C (140°F). In the machine shop, VSE boss, Tim Camp, set about surface grinding. It's vital to ensure that the timing case cover is in place at this point, otherwise there'll be a 'step' when the cylinder head is replaced.*

▶ *Even with all this technology, the only foolproof way to ensure that the surface is exactly level is to check it with a steel straight-edge and a set of feeler gauges, as Tim is doing here. In this case, the head required just 4 thou off the top – it's common to find that 8 thou is needed.*

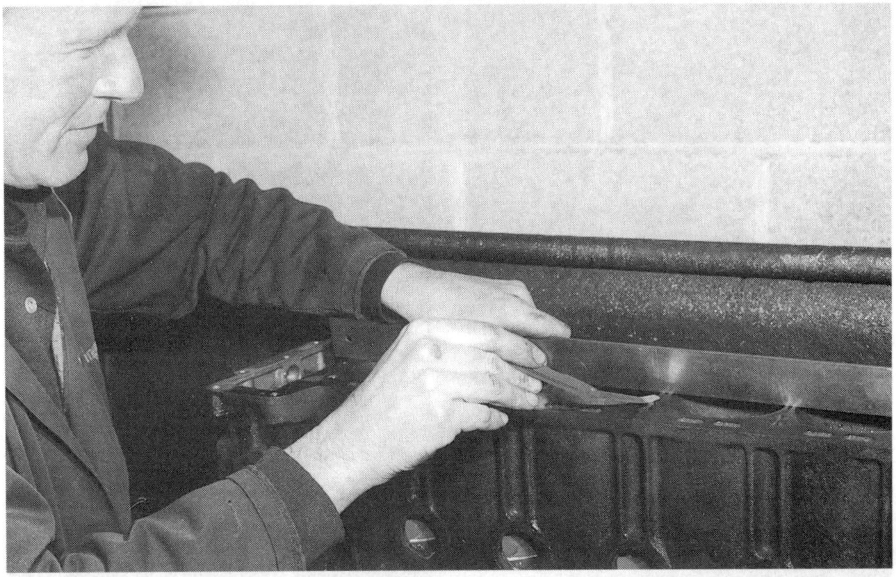

▶ *Tim checked the bores carefully for signs of taper, ovality, scoring and scratches. A good idea of what the bores will be like can be gained by knowing how the engine performed when in the car. Excessive oil consumption and lots of blue smoke indicates worn bores. He used a Vernier calliper to measure the diameter at the top of the bore and then again at the bottom, where there is no wear. If the difference is more than 0.006in (0.15mm) then a rebore with oversize pistons will be required – and this was the case with this engine. Tim set up the cutting head and wound it down into each bore in turn. It was set to 20 thou oversize, because 10 thou oversize pistons are not available. Note that reboring can only be carried out up to 30 thou on this 4.2-litre fuel-injection block – if any more is required, it means that the liners must be pressed out and new ones fitted in order to start all over again.*

▶ *More wisdom from Tim tells us that later 4.2-litre engines had a change of waterway position between the bores, and thus the maximum rebore was reduced to +0.020in. In early 4.2-litre engines (as shown here) the block has waterways between the bores behind the liners. There's a very good chance of small cracks between the bores. If left, the head gasket will blow, with possible cracking of the liner.*

◀ *The bores were finished off by Russ, who used a flex, or bottle brush, honer. After oiling the cylinder bores, the brush was connected to an electric drill and moved up and down the bore. Note that it is important NOT to remove the brush while the drill is rotating. This honing produces a crosshatch finish and a good key for piston rings to work in and it is essential that newly machined bores are honed before fitting the pistons. With the honing complete, he chamfered the top edges of the cylinder bores carefully with a small file so that the rings wouldn't foul when the pistons were installed. The block was then steam cleaned. Once dried, it was wire-brushed and painted with a heat-resistant synthetic black enamel paint.*

▲ *Where this occurs, the best solution is to remove the liners and machine the block to take a flanged liner (often referred to as a 'top-hat' liner, for obvious reasons). This allows for the sealing ring of the head gasket to seat on to the top of the liner, ensuring a 100 per cent seal. Clearly, with the liners removed, the waterways should be thoroughly cleaned out – they're bound to be silted-up!*

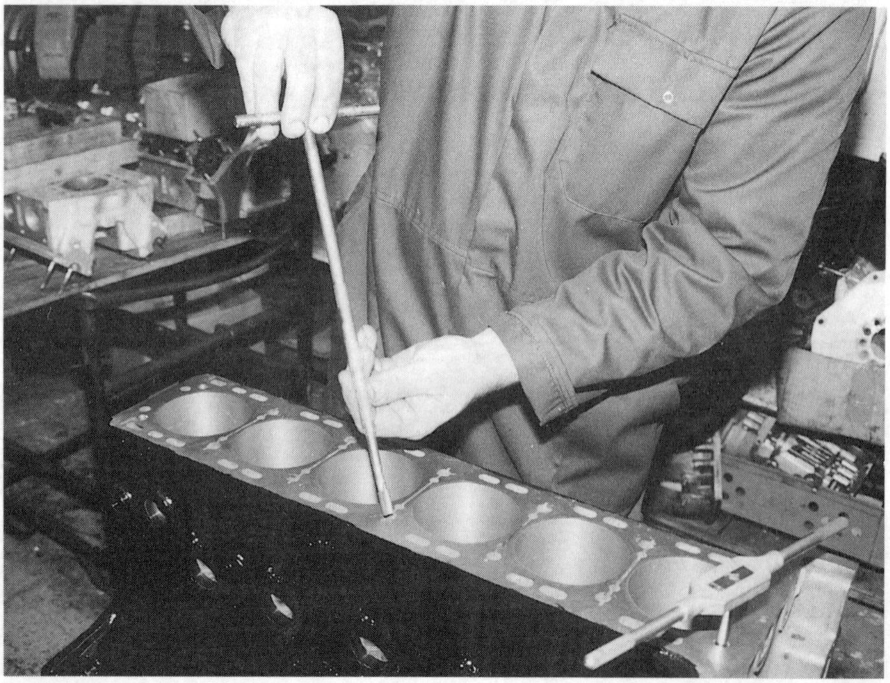

◀ *Malcolm was in charge of engine reassembly, and he took the trouble to run a tap through all the threads on the cylinder block. Yet more attention to detail that is typical of the business as a whole and essential when you're building engines that go all over the world.*

CRANK STRIP, EXAMINATION, BALANCE AND REBUILD

◀ *The sprockets and worm gear on the end of the crank are usually OK. On earlier models there is sometimes an oil thrower plate which can often be damaged and has to be replaced. The crank has six Allen-headed recess plugs, but each has a peen mark which has to be drilled out before the plugs can be removed. On some older XK units, these plugs will probably need drilling out because they've been there so long.*

▶ *There's almost certainly going to be plenty of messy silt behind each plug, which means that the crankshaft oilway is equally sludged-up. Obviously, this reduction in lubrication is going to do the crank a power of no good. Given that the main reason for this (common) XK problem is simply not changing the oil often enough, there's really no excuse – moreover, there are plenty of reasons to make regular oil changes a part of your life, most of them financial! VSE always chase out the plug threads before going on to the next stage.*

◀ *VSE has complete facilities for all aspects of engine building, including full crankshaft testing kit – something you simply can't do at home, folks. Here, Tim is checking 'our' crank for cracks. The crank is magnetised by inducing a current through it. He then sprays a fluid over it which contains dyed magnetic particles which show up in ultraviolet light (hand-held in this photo). In the darkness of the covered booth, a crack will show up as a bright yellow line. Cracks usually occur where the journal joins the webs of the crank, although this is rare on the 4.2-litre cranks.*

◀ *Tim used a micrometer to check the journals for size and ovality. 'Our' crank was within the specified limits, so he ground it as required. He ground the first journal and then the others to match it, repeating the process for both big end and mains surfaces. Tim found that the only real way to ensure total quality was to install the necessary machinery and carry out the work on the premises.*

Technical note: According to Tim, Series 3 XJ6s have a crank that is surface-hardened by the 'Sursurf' process (from engine No 8L147650 4.2-litre and 7M4796 3.4-litre). In theory these cranks can't be reground, as it removes the hard surface. However, it is possible to regrind and retreat the crank to restore its hard surface – a job for the professionals. He checked and rechecked the journals at regular intervals, and when they were correct …

▶ *… he transferred the crank to yet another complex machine for balancing. Here, the crank was mounted on a pair of rollers with vibration sensors underneath which transmit the imbalance to the in-built computer. Imbalances show up as a digital display, so Tim knows exactly where to remove metal from. Once balanced, Tim added the flywheel drive plate and front damper so that everything was balanced to itself. With the crank perfectly balanced and cleaned-up, new plugs were fitted and new peens made using a suitable punch. Standard plugs only have one peen but Malcolm, ever the perfectionist, prefers to make two – just in case.*

◀ *Know your cranks! The 4.2-litre cranks were fitted with quite large recess plugs, but earlier models had small ones, the larger plug cranks being better and more efficient. The large plug crank seen on the right in this photo is a special racing crank which, despite being for a 3.8-litre E-type racer, has bigger plugs and is, naturally, considerably stronger than the standard item. A normal 3.8-litre crank is shown on the left.*

▲ *For an even smoother engine, you could fit a fluid harmonic crankshaft damper to replace the original. It's of use on both standard engines and those that have been breathed upon.* (Courtesy Rob Beere Racing Services)

THE PISTON AND CONNECTING ROD ASSEMBLIES

▲ *Note that the caps were replaced on the con rods as the pistons were removed from the engine so they could be put back together as matched sets. Malcolm started by removing the circlips from the piston gudgeon pin (little end). He threw away the little ends, and then checked the rods for condition. If the stamping inside the big end (showing the rod designation) is still clearly visible, then the bearing shell hasn't been loose and wriggling about causing wear. If the little end bronze bearing can be pushed out by hand, it's very bad news and very worn, but also extremely rare.*

▲ *The con rods went to Russ for cleaning and came back looking like new. Here, they are lined up ready to be assembled with new little ends, circlips, pistons rings and Hepolite pistons.*

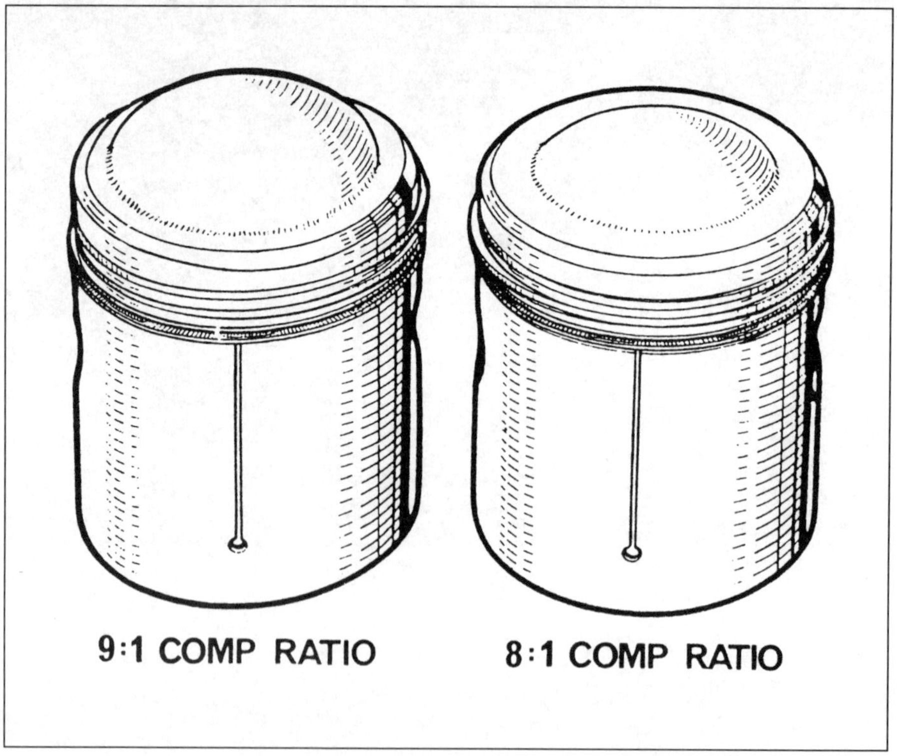

▲ *Almost all XK engines were offered as standard with a choice of compression ratios (anything from 7.0:1 to 9.0:1) with power outputs usually around 10bhp apart. These differences were down to different pistons being used, the high compression ratio pistons having a higher centre 'dome'. This diagram shows the difference between two piston sizes available for the E-type, the one on the left being 9.0:1 and on the right 8.0:1.*

▶ *As ever, the more power you expect from your engine, the more specialised the parts have to be. Seen here is a collection of three pistons from the Rob Beere Racing Services workshop. All are for the 3.8-litre E-type. On the left is a standard 8.0:1 piston made by AE; in the centre is a wide-angle D-type Cosworth piston and on the right is a narrow-angle piston designed by RBRS and produced by Cosworth. The second two are designed to handle hefty power outputs and can be anything up to 12.0:1 compression ratio (RBRS would expect around 11.5:1 to be the norm).*

◀ *The narrow-angle pistons can be seen doing what they do best in this full race E-type engine.* (Courtesy Rob Beere Racing Services)

▶ *Early XK con rods were drilled up the centre (as can be seen on the rod to the left of this photo), whilst later rods did not have the hole and, in addition, had a wider blade. The rod on the right is a Carrillo rod – a seriously tough cookie with a price to match (at the time of writing, even the nuts cost £300 a set!) However, according to RBRSs Carl Taylor, they've fitted Carrillo rods to XK units with the most outrageous power and torque outputs and have yet to have a problem with these rods – regardless of how 'enthusiastically' they were driven. Expensive, yes, but it seems you get what you pay for.*

◀ *The bronze bush in the little end should have a diameter of 0.875in + 0.002in (22.22mm + 0.005mm). Anything more than that means it will have to be replaced, although Malcolm usually replaces them as a matter of course. Getting the bush out of the con rod is virtually impossible without the use of a hydraulic press. Here, Malcolm is using a suitably-sized drift and taking care to ensure that it is being pushed out 'square'.*

▶ *The bronze bushes have an oil hole half way along their length and it's vital that these holes line-up exactly with the corresponding holes in the con rods during assembly.*

▶ *The bush has to be reamed out for an exact fit with the new little end. For DIY purposes, this would probably be done in a vice, but great care must be taken to ream absolutely straight. Malcolm devised this impressive device, comprising a piece of round bar with the same diameter as the crank and two old con rods. Having pressed the new bush into the con rod, it goes in the middle of the device and the reamer is pushed through from one side to the other.*

◀ *He reamed a little and checked, then reamed some more and checked again until the fit was almost right. He then used a little emery paper and a lot of skill to bring the size just right.*

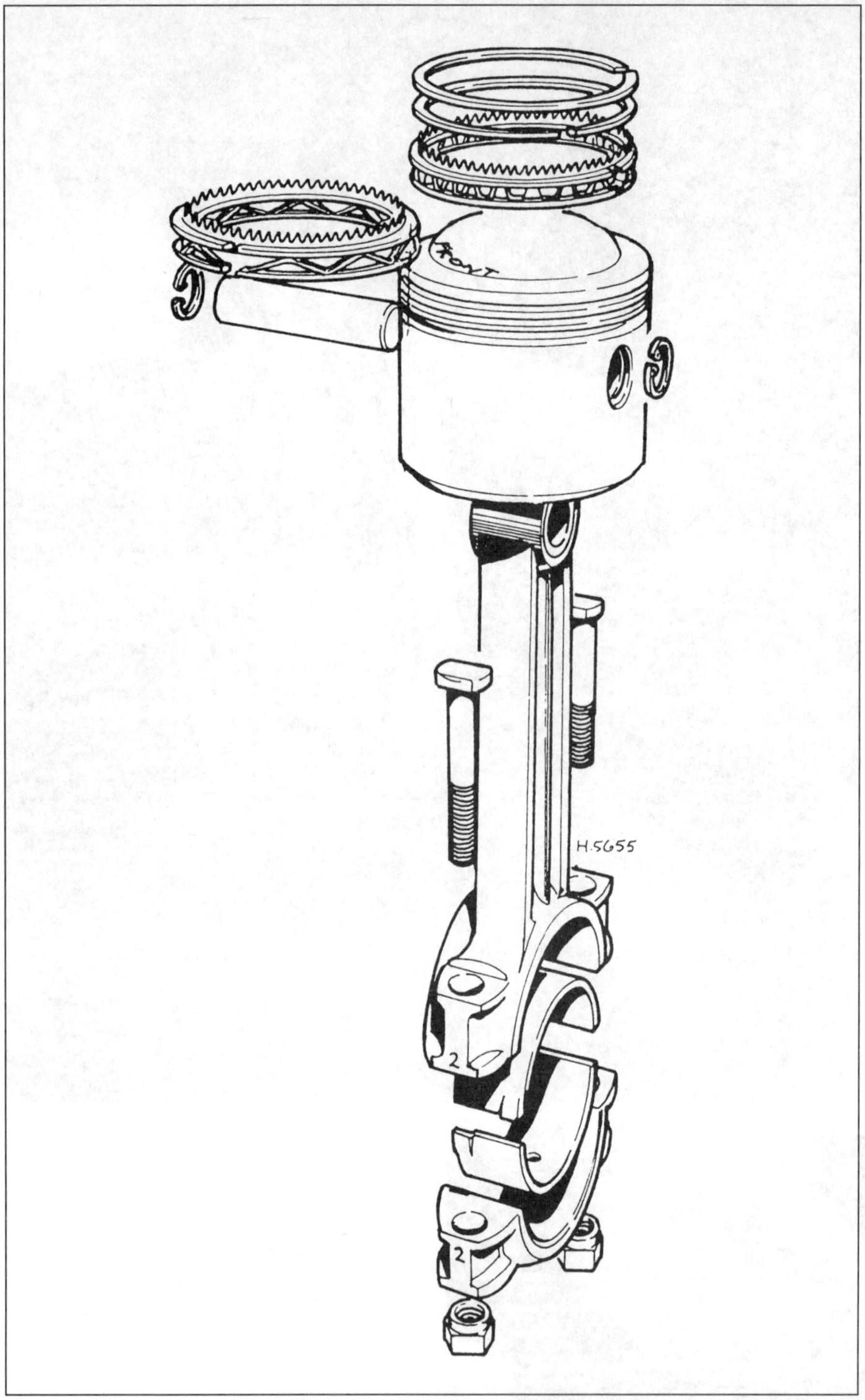

▲ *This exploded diagram is of the piston and con rod assembly for the 4.2-litre XJ6 engine – both types of oil control ring are shown.*

▲ *For the 4.2-litre engine there are three rings; the lower is the oil control ring, which comes in two parts and when assembled must not overlap – it should abut perfectly. The other two rings are scraper/compression rings which are tapered, so it's important that they are fitted with the word 'top' uppermost. Malcolm weighed the piston assemblies and balanced them by removing small amounts of metal from inside the lower skirt area. Naturally, they have to be balanced to the lightest piston and the same principle applies to the con rods.*

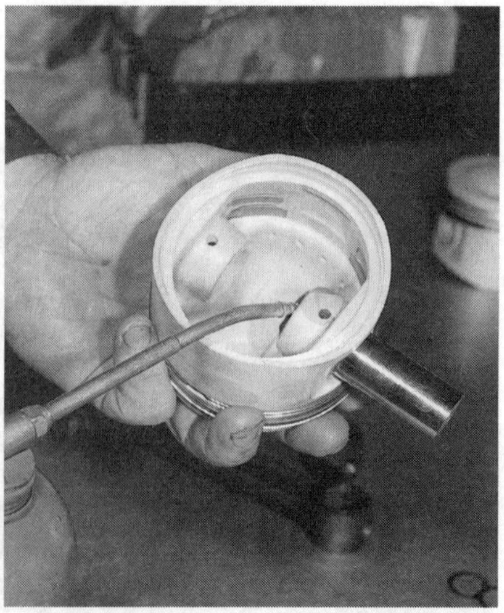

▲ *He then fitted one of the little end circlips and oiled the little end ready for it to be pushed into the piston. Once it was in this position, he applied more oil to the oil hole to help lubricate it and …*

▲ … the con rod bronze bush as the little end was pushed fully home, at which point the second circlip was fitted. Missing off a little end circlip is a very common DIY affliction and a very simple way to wreck your new engine.

▲ Malcolm also had other rebuilds in progress – these are the piston assemblies for a later 3.4-litre XK unit. Spot the difference? There's an extra piston ring. It's important to get exactly the right pistons/rings for your engine, as they changed a great deal over the years.

▼ These are the complete piston and con-rod assemblies for an XK 120.

◀ *This photo shows the various types of con rod used throughout the XK's life, running latest to earliest, left to right. Early engines used a pressure-fed con rod, which had a hole running up its centre (ie the rod was hollow, as seen here on the right). Later versions kept the hole up the middle of the forging but not at the bottom, and latterly there were no holes or galleries. For a fast, road-use engine it's generally advisable to move from the centre hole rods to the no-hole, wide blade versions.*

◀ *The other major modification to the con rods was the change of fastener – early models relied on the good old castle nut and a split pin to secure them in the big ends. But, latterly, this was changed to stretch bolts and bi-hex nuts, which were easier to use and far more reliable. VSE always fit the latter in a rebuild.*

Chapter 13

Timing Gear – Dismantling, Preparation and Reassembly

The twin-chain timing gear assemblies are complex, as seen during the dismantling procedure, and some thought and care is needed when it comes to reassembly. Always, but *always* replace the chains during the rebuild process. If either chain stretches (and both will in time) then it will become too slack and the tensioner block can come out. The results of this are obvious, messy and extremely expensive. It is possible to swap the top chain on most XK units with the engine *in situ* – although it's not a task you'd look forward to. Swapping the lower chain is an engine-out task. This is yet one more reason to renew both chains having got the engine to this stage.

▲ *Early XK engines (up to 1954) used this kind of blade/spring type tensioner, which was not the most successful in the world. It was replaced …* (Courtesy VSE Engineering)

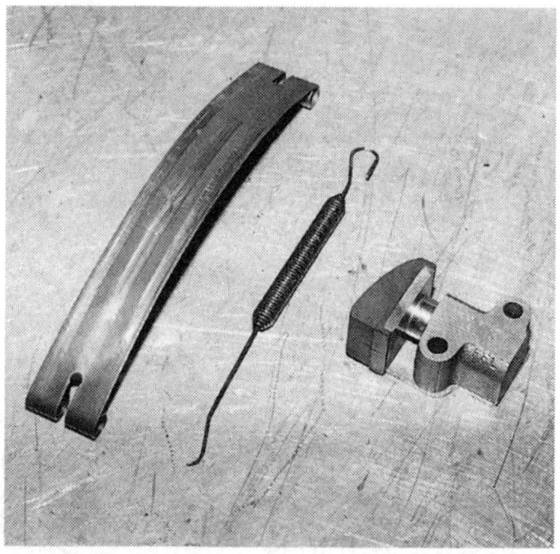

▲ *… by a hydraulic version, seen here to the right of the original type.* (Courtesy VSE Engineering)

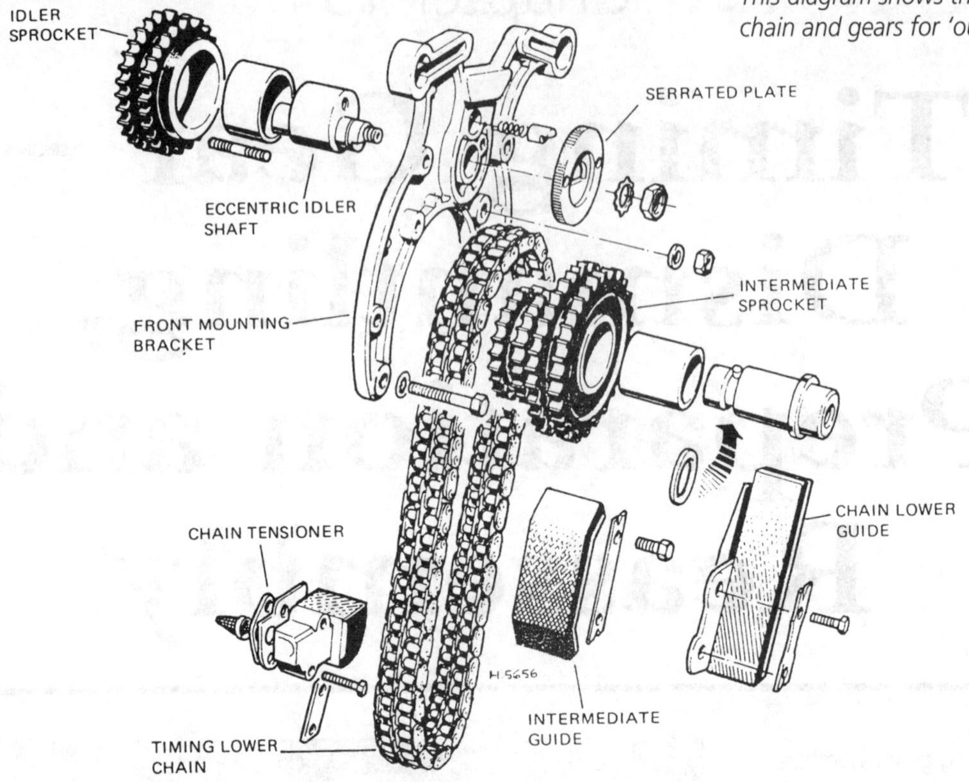

This diagram shows the lower timing chain and gears for 'our' XJ6 and …

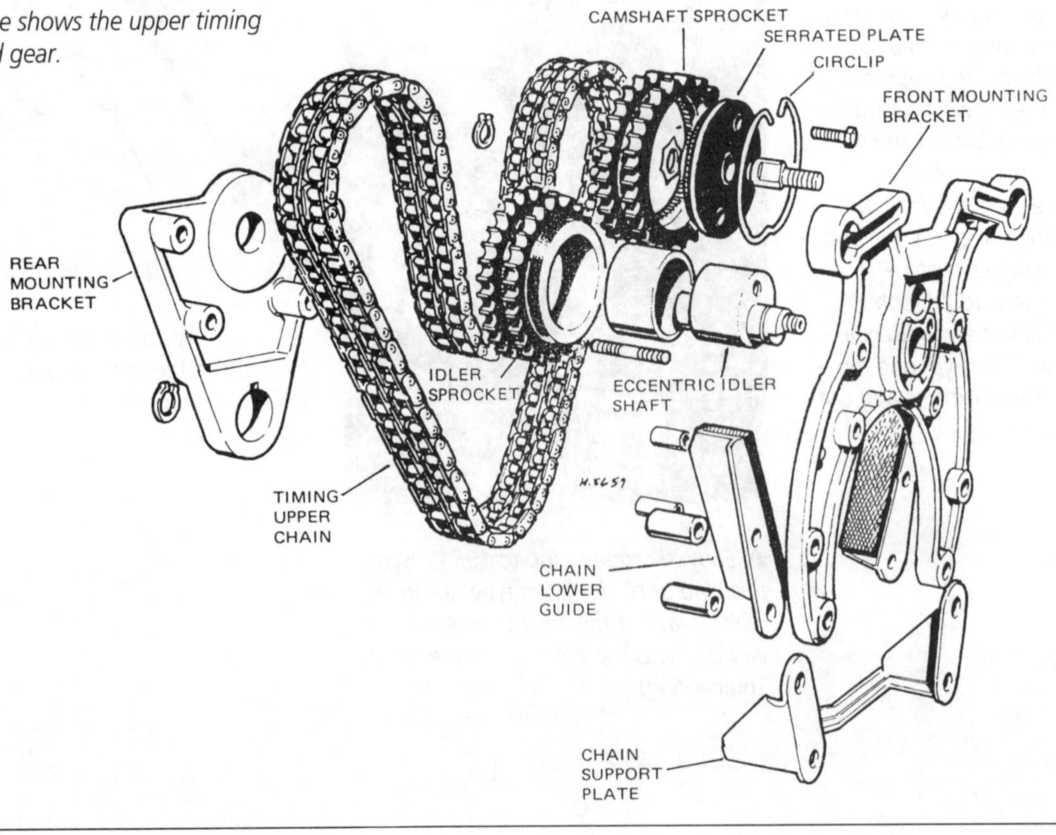

… this one shows the upper timing chain and gear.

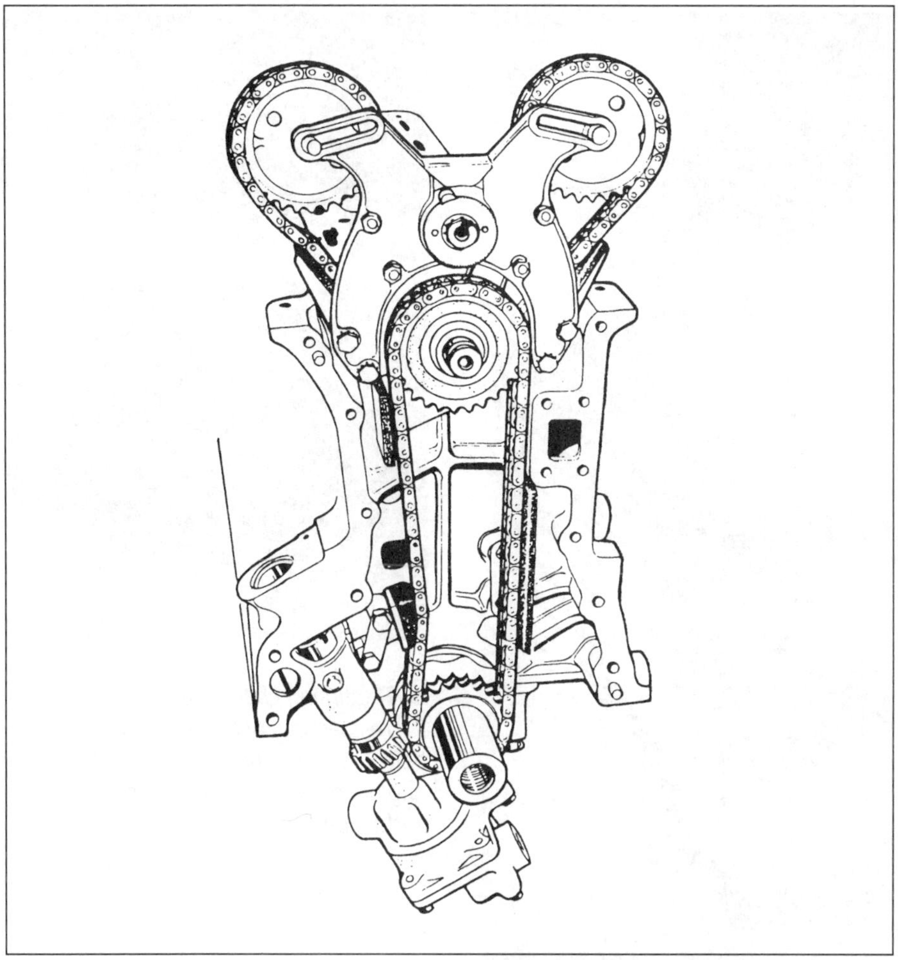

◀ *This shows how the completed timing gear should look on a 3.8-litre E-type unit. Note that the cam sprockets have only two holes – later engines (such as the XJ6, shown for most of the work in this section) have four.*

▶ *After removing the assembly from the engine, Malcolm took off the elastic band that held the cam sprockets together and started the dismantling procedure. With the circlips removed, he was able to tap the two timing chain brackets apart.*

▲ The timing chain components ready for reassembly, with new chains, nuts, bolts, washers and circlips, but the original mounting brackets, sprockets and guides.

▲ Sprockets rarely need replacement, but obviously need to be checked as a matter of course. Ensure that the teeth are not hooked and that none is damaged. If you're unsure, it's far cheaper to replace them now than have to tear down the engine again at a later date. In this case, they were cleaned up by Russ along with other timing gear hardware in the Clarke washer with Safety Kleen before being buffed up on the wheel – note safety goggles and gloves.

▶ Compare with a very early XK 120 set-up. There are few differences, the most obvious being the inclusion of the spring-type blade tensioner and the twin (rather than four) hole sprockets. (Courtesy VSE Engineering)

▶ *The camshaft sprockets have serrated plates in their centres that are held in place by large circlips.*

▶ *The pin was lightly oiled before being inserted into the intermediate sprocket. The bush in the centre of this sprocket and the idler sprocket can wear. Malcolm fitted new bearings into the two sprockets. Test by inserting the shaft and rocking it side-to-side – if you can feel any play, then the bush(es) will have to be replaced.*

◀ *The assembly was pushed into position on the front mounting bracket and held by a circlip. The eccentric idler shaft was given a coating of engine oil before being put into position.*

◀ The longer of the two new Rolon chains was looped around the sprockets, then the shorter one added.

▶ Malcolm put the front mounting bracket in place and, before tightening the nuts (which fit on to studs in the bracket), used an elastic band across the camshaft sprockets to prevent them going for an unauthorised 'walkabout'. The lubricating spring and plunger were placed in position and …

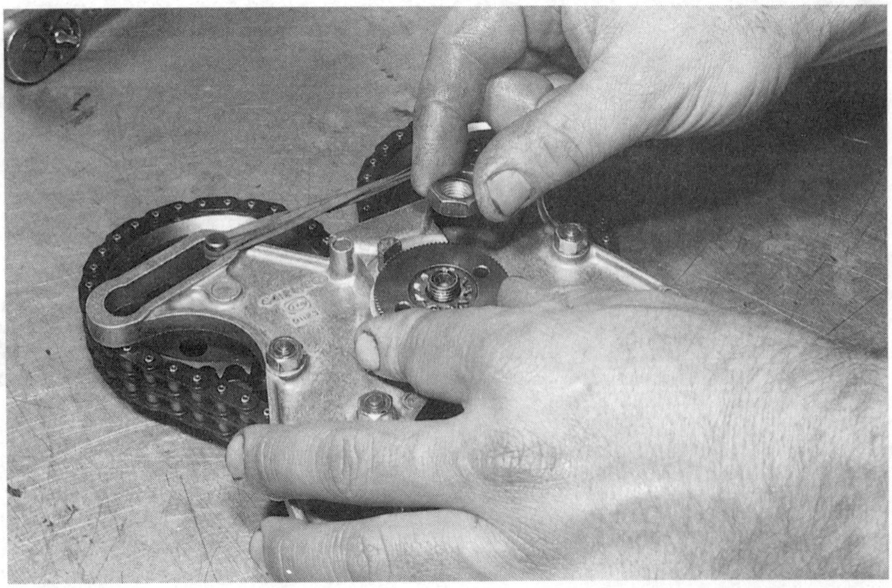

◀ … the eccentric idler shaft fitted and secured by a washer and nut. The lower chain tensioners were put in place. The 4.2-litre engine also has a bridging plate.

▲ *Two completed timing chain assemblies; the one on the left is for 'our' 4.2-litre engine and the one on the right is from a 3.8-litre XK unit, for direct comparison.*

Chapter 14

The Cooling System

All XK engines are particularly intolerant of overheating, meaning that the cooling system must always be in top condition. Unfortunately, many owners neglect it as they would on a Ford Fiesta, with equally unfortunate results. Cylinder head problems are common and, obviously, excessive heat on a regular basis will wreck the engine altogether.

As such, it is VITAL to keep the coolant fresh – change it at least every two years, or annually if you really love your engine. Remember that the less you use your car, the more attention you must pay to your coolant. A 50/50 mix of anti-freeze and water is required, although the term anti-freeze is a misnomer and, indeed, many companies are now developing new terms for it. Its role in your car's engine life is twofold – first, it prevents the coolant in the system from freezing in winter, and second, it raises the boiling point of the coolant in summer, reducing the likelihood of overheating. Moreover, in the case of the XK, it also helps prevent the rusty sludging so prevalent in older and/or neglected units.

Regardless of age or type of XK engine, the cooling system is fairly simple. The water circulation is assisted by an impeller-type pump mounted on the front of the engine. The system is pressurised and thermostatically controlled. The pressure is controlled by the radiator (latterly, the header tank) cap, and periodic checks on the cap are essential if efficient cooling is to be achieved. If the cap leaks, then the pressure is relaxed and, as the correct pressure should raise the boiling point of the coolant by more than six degrees, you can see its importance.

Water is brought from the base of the radiator block via the water pump, through the cylinder block and the cylinder head passages, and is returned to the radiator header tank by way of the inlet manifold water rail (and thermostat). The water is cooled in its passage from the header tank through the radiator matrix to the base of the radiator block. In some models, to prevent air locks, an air bleed valve is fitted in the air vent aperture in the water outlet elbow.

DRAINING THE COOLING SYSTEM

IMPORTANT NOTE: Some Jaguars are fitted with air conditioning, and the system MUST be discharged before removing the radiator. On some models (the S3 XJ6, for example), it is not strictly necessary as long as the condenser and receiver/drier can be moved out of the way without placing undue strain on the hoses. However, great care must be taken when dealing with any part of the air conditioning system.

SAFETY: To avoid personal injury, only drain the coolant when the engine is stone cold. Draining a hot engine also runs a risk of cylinder head distortion and lots of consequent expense!

FLUSHING THE COOLING SYSTEM

Remove the radiator cap (cold engine) and look inside. Run your finger around the inner lip of the radiator and if it comes out covered in slimy brown sludge, flushing is long overdue.

FLUSHING THE RADIATOR

Remove the radiator from the car and turn it upside down. Insert a standard hose pipe (ie NOT a high-pressure washer!) into the bottom hose hole. Make sure that water only comes from the top hose hole and is not leaking from elsewhere. You'll see the brown sludge coming out at first, but then you should get clear water.

FLUSHING THE ENGINE

Open or remove the drain tap from the engine block and also the thermostat. Insert the hose into the thermostat hole and flush through until clear water runs from the drain tap.

FAN AND WATER PUMP

Early XK units used a multi-bladed fan at the front of the engine (fitted to the spindle of the water pump) to keep it cool. The first XK fans had four blades, but this was ultimately increased to 12. It was turned by a belt from the crank pulley which also passed around the dynamo pulley. It worked all the time the engine was running, which meant that at some times (notably at high speed or cold starting/ tickover) the engine could be overcooled.

The viscous fan features a fluid driven coupling which slips at between 1500/1900rpm. The centre finned assembly to which the fan is bolted is called a Torquatrol drive unit and it is

▶ *Before draining the system, set the heater controls at 'hot' and use a large receptacle to catch the coolant. Where remote rod-type drain taps are fitted, it's a fair bet that they will have rusted solid many years ago. Soaking in releasing agent can help, but not always. If it doesn't, don't force it, as you'll only make things worse. There will also be a drain at the rear of the cylinder block – some models have a drain tap, others a drain plug.*

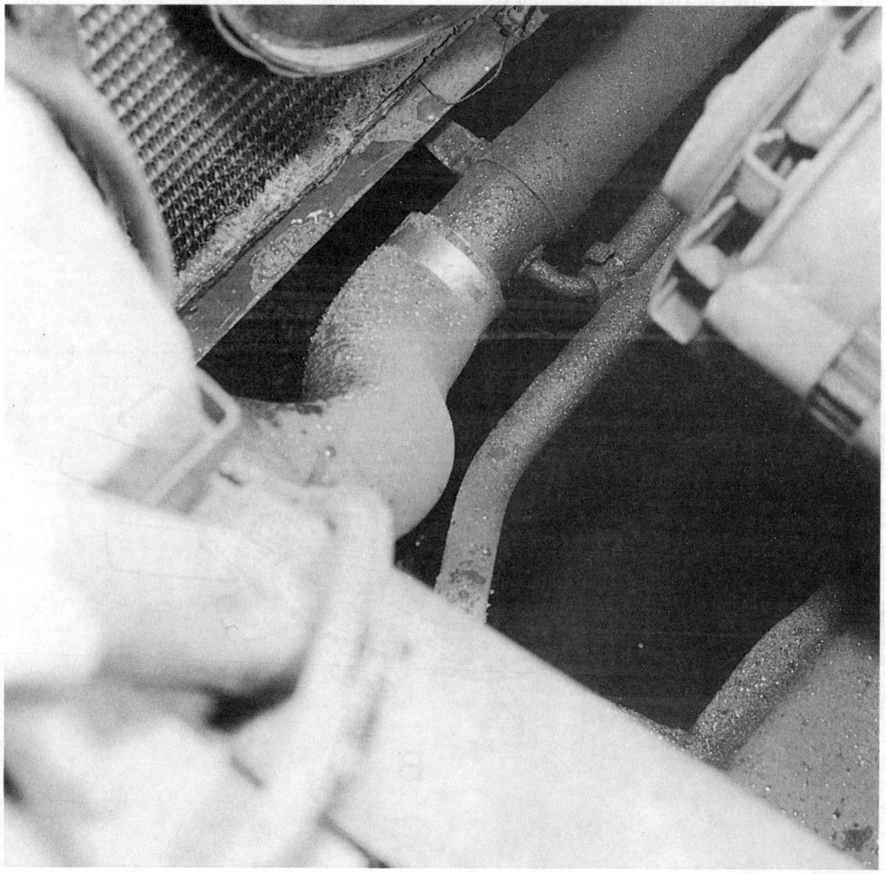

▼ *Unscrew and remove the header tank/expansion chamber filler caps. The system is pressurised, so this is another reason for NOT working on the cooling system when it is anything other than cold. The expansion/header tanks can only be emptied by removing them from the car. Being an automatic transmission XJ6, this particular model had a fluid cooler in the centre section of the radiator bottom hose, as seen here.*

▲ *The rear tap was opened and (as the radiator tap had seized) we removed the plug from the bottom water pipe and drained the coolant from there also. Never re-use anti-freeze, but make sure you dispose of it responsibly in the same way as used oil (see Chapter 6).*

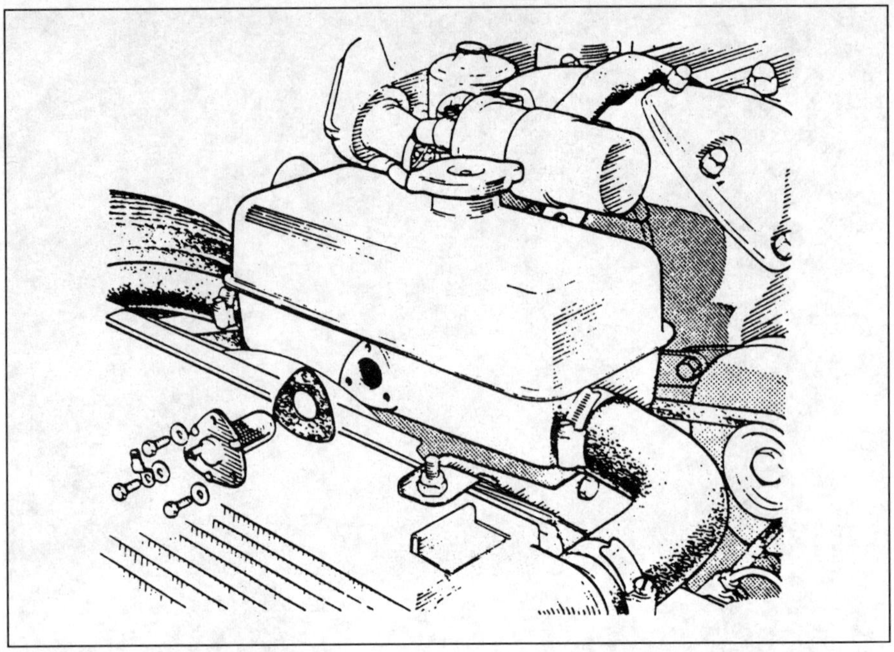

◀ *Gradually, more modern fan systems were installed. Depending on the car you have, it will be equipped with either an electric fan, triggered by a temperature sensor, or a viscous version. Typically, electric fans are set to cut-in to provide extra cooling for the radiator at around 80°C (176°F) and switch off automatically at around 72°C (162°F). This diagram shows the positioning of the thermo switch in the header tank of an E-type.*

this which slips as the engine speed rises, thus effectively removing the drive to the fan. It is fitted to early water pump pulleys by four bolts, and to later pulleys by a single, centrally-positioned bolt. This, too, can promote overcooling in some instances at idling, etc – and some owners, more intent on practicality than originality, remove the viscous fan and replace it with an electric version, typically from companies such as Kenlowe and Pacet. Clearly, it's important to take professional advice here, as this kind of move will depend very much on the car you have – owners of early E-types know only too well the true joys of a sports car which overheats in heavy traffic!

▼ *This diagram shows the fan and fluid couplings for XJ6 models: (a) early cars and (b) later cars, and …*

▲ ... *this photo shows the complete late model assembly in the flesh.*

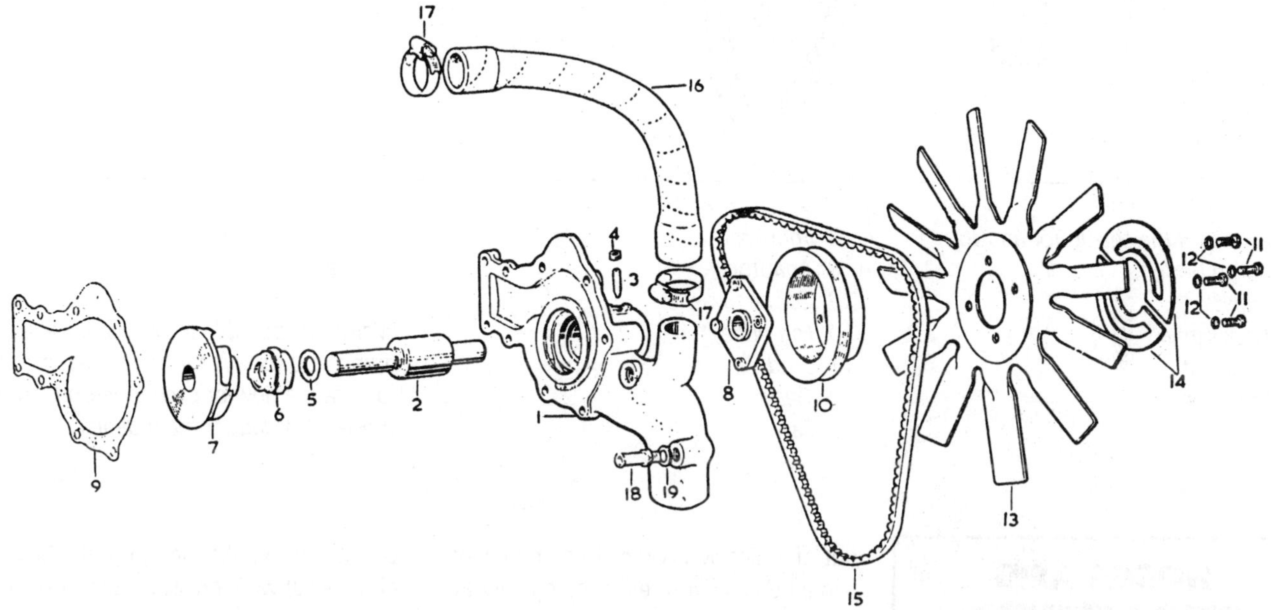

Though water pumps have improved in overall efficiency over the years, the basic layout and design of them and the associated fan and pulley components has remained more or less static. This is the layout for a Mk 2 car.

1. Pump body
2. Spindle and bearing assembly
3. Allen headed screw
4. Lock nut
5. Thrower
6. Seal
7. Impeller
8. Pulley carrier
9. Gasket
10. Fan pulley
11. Setscrew
12. Shakeproof washer
13. Fan
14. Balance piece
15. Fan belt
16. By-pass water hose
17. Clip
18. Adaptor for heater return pipe
19. Copper washer

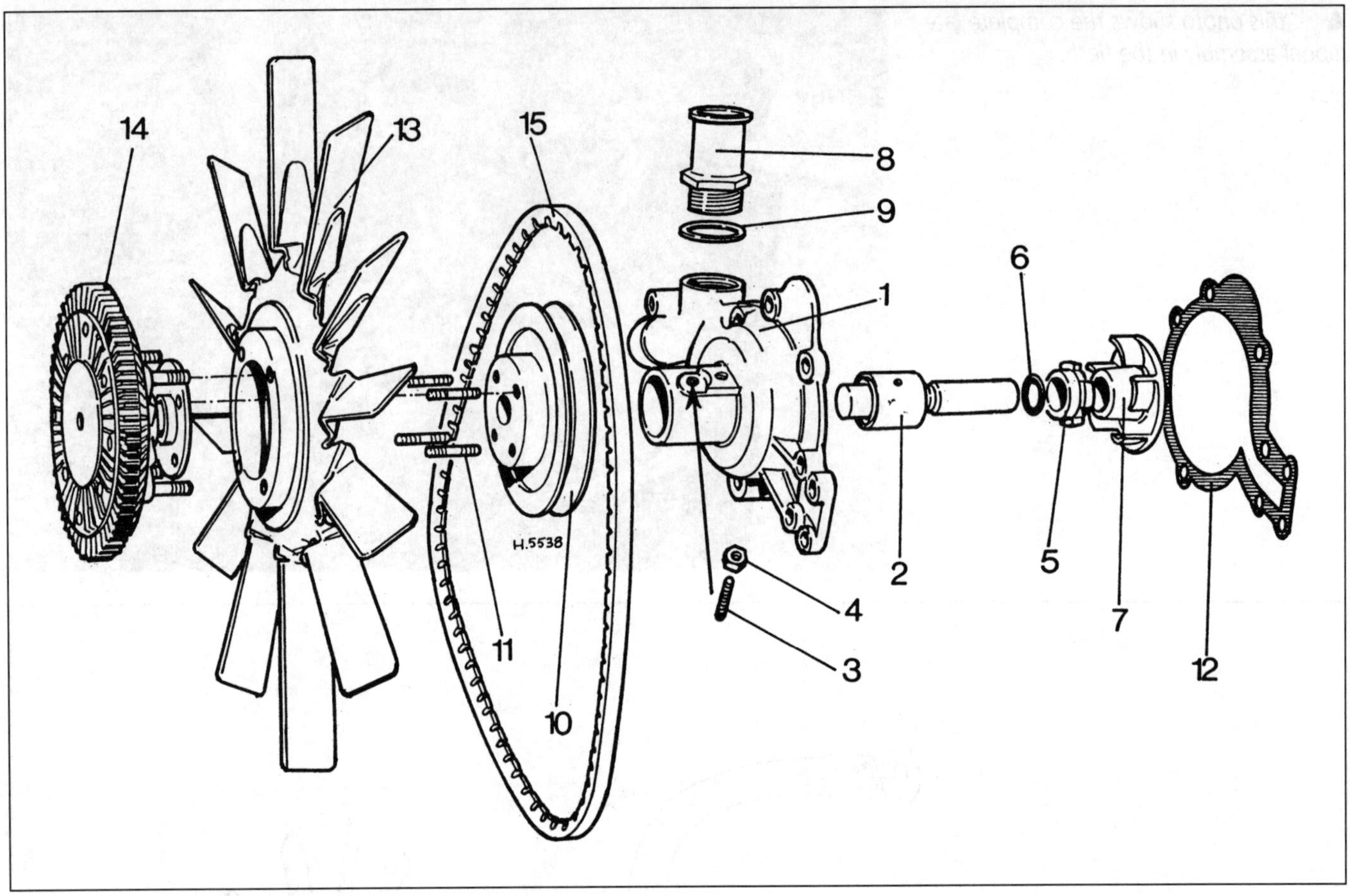

▲ *And this is the set-up for an XJ6:*
1. *Pump body*
2. *Spindle*
3. *Lock screw*
4. *Lock nut*
5. *Thrower*
6. *Seal*
7. *Impeller*
8. *Adaptor*
9. *Copper washer*
10. *Fan pulley*
11. *Stud*
12. *Gasket*
13. *Fan*
14. *Fluid coupling*
15. *Belt*

Whilst it is possible to overhaul the water pump, it is usually far more cost-efficient to obtain a new or exchange unit from a specialist or Jaguar main dealer.

HOSES AND REPLACEMENT

All cooling hoses should be regarded as suspect unless you can be absolutely sure that they are fairly new. Even so, inspect all the hoses carefully on a regular basis – bend them back harshly (cold engine) and look for even the slightest sign of a hairline crack or perishing. If you find any, replace the hose. A small crack may not look bad, but eventually it will give up under the pressure and heat to become a 3in split, out of which will come litres of coolant – in a hurry! If you're on the motorway when it happens, you could have an engine in kit form by the time you can stop. Check the jubilee or spring wire clips securing the hoses at either end for tightness.

RADIATOR RUDIMENTS

The radiator is a core of thin-walled, metal tubes with fins attached. Early XK-engined Jaguar radiators were brass with copper fins, whilst the tanks (the bits at either end) were brass. In common with most manufacturers, modern Jaguar radiators tend to be aluminium cores (tubes and fins) with plastic tanks, although some brass/copper cores can have plastic tanks.

As the car moves along, the cool air flow passes over the tubes and cools the water inside them. The fins also help dissipate the heat. The hot water from the engine goes into the radiator at the top, and by the time it reaches the bottom it is much cooler and ready to re-enter the system and start its journey once more.

INSPECTION

Take some time to inspect your radiator thoroughly, paying particular attention to the condition of the cooling fins and checking for obvious signs of leaks. If you have been using the right mixture of anti-freeze, there will be a tell-tale trail of dried blue/green coolant showing where it has been escaping. If you've been lax, it will be a dirty brown (ie rusty) trail.

The radiator is almost universally overlooked until it is seriously clogged up or rotten (or both). The trouble is that it may have been under-cooling for some time before it is evident to the driver, which means that the engine could have been running too hot for many thousands of miles. The relatively small expense of replacing or repairing the radiator can often offset the inevitably large expense of a major engine rebuild. And, of course, a failing radiator places extra strain on other cooling system components, notably the hoses.

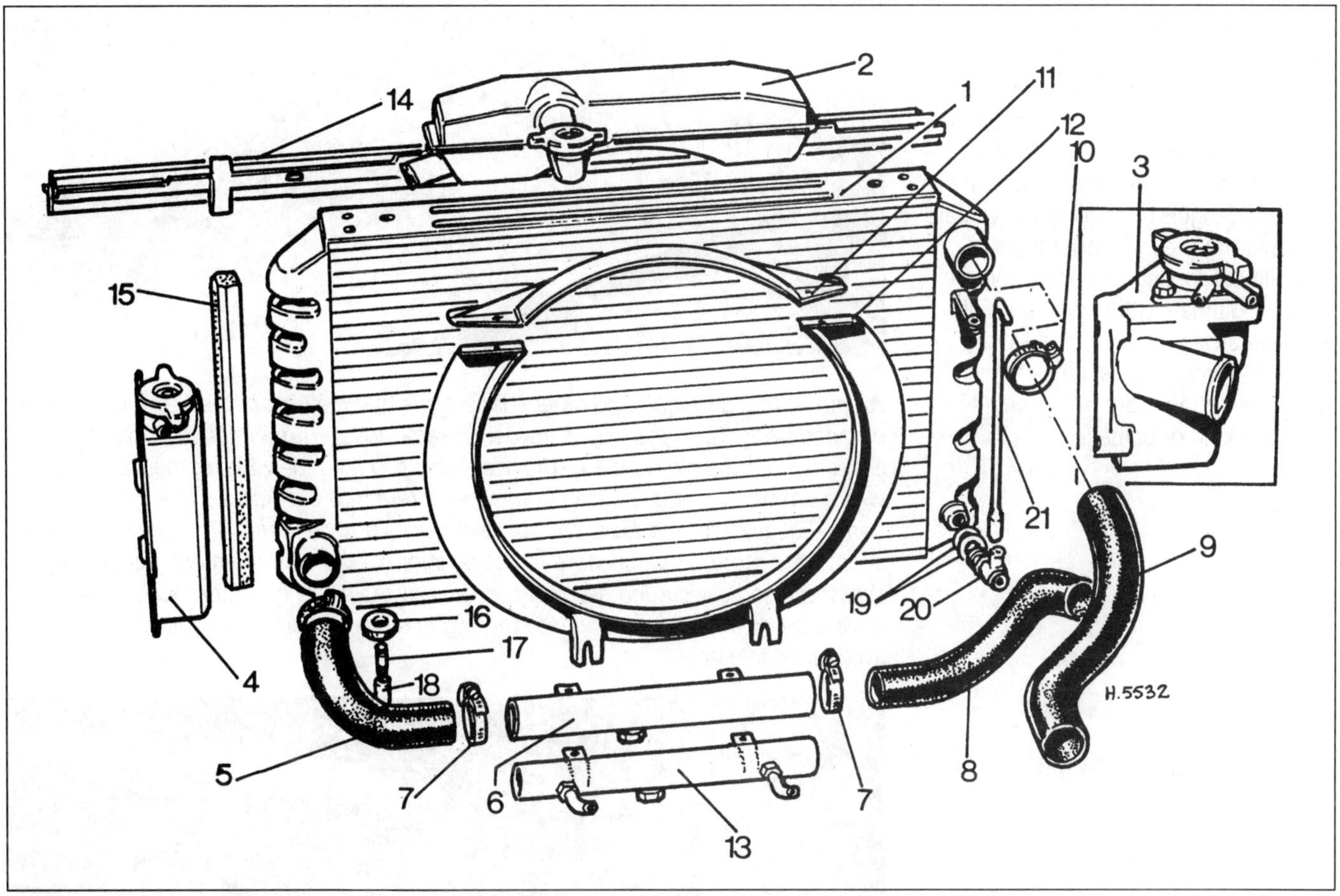

▲ *This diagram shows the basic radiator component layout on the XJ6. All XJ6 models have crossflow radiators, so-called because the tanks are on either end of the core, rather than the more usual top and bottom. However, the way they work is exactly the same.*

1. *Matrix*
2. *Header tank (smaller engines)*
3. *Header tank (larger engines)*
4. *Expansion tank (larger engines)*
5. *Radiator hose*
6. *Bottom water pipe (manual box)*
7. *Hose clip*
8. *Radiator*
9. *Radiator top hose*
10. *Hose clip*
11. *Radiator cowl (top section)*
12. *Radiator cowl (bottom section)*
13. *Bottom water pipe (auto transmission)*
14. *Upper cross panel*
15. *Sealing strip*
16. *Mounting rubber*
17. *Stud*
18. *Distance piece*
19. *Fibre washer*
20. *Drain tap*
21. *Tap remote control*

REMOVING THE RADIATOR

Removing the radiator for examination, repair or renovation is a job which can be easy or hard depending on the car you have. The XK series sports cars and, say, the Mk 2 models are somewhat cramped. The XJ6, seen here, and particularly the E-type have rather more room under the bonnet which makes life somewhat easier. In most cases, it's ideal to remove the bonnet altogether. One of the main dangers (assuming that you are not replacing the radiator) is the possibility of fouling and damaging the fins and, ultimately, the core tubes. If you have a car with limited radiator access, it's wise to enlist a helper to assist in wiggling it out of its home to avoid this kind of damage.

▲ *Ensure that all wiring and other connections that may get in the way are removed or taped back out of the way. Remove all water hoses – if the radiator is old it can be expected that the old rubber of the hoses sticks to the pipes. Easing them slightly with a screwdriver and spraying in some WD40 usually encourages them to leave home. Equally, you can expect a Jubilee clip or two to be well-rusted, so use releasing agent once more. You may have to use a junior hacksaw on the more recalcitrant. Whilst accidentally cutting hose isn't the end of the world (you'll probably have to replace a few anyway), cutting into the wiring or metalwork of the pipe is to be avoided at all costs.*

▲ *With everything removed, the radiator can be lifted clear. The XJ6 in this case had had the bonnet removed altogether – quite a simple operation as long as there's a helper around to take some of the weight! In many XK-engined Jaguars, it is nowhere near as simple, though that helper still comes in handy.*

RADIATOR RENOVATION

When your radiator has passed its sell-by date, there's not much to be done on a DIY basis. However, you don't just have to scrap the radiator, because Northampton Autorads has a wealth of experience in repairing and making radiators and associated parts stretching back over 30 years to when Don Goodwin first started the business. It's grown since then, and now his son Bob is involved along with an enthusiastic staff headed up by workshop manager David Chater.

Autorads is heavily involved with the Jaguar scene, dealing with over 1,000 rebuilds/repairs every year for just about any breed of 'cat' you can name, going all the way back to the SS series! As a further measure of quality, since 1975, Aston Martin have fitted Autorads radiators to every Aston Martin that has left Newport Pagnell! Perhaps it's no surprise that there's now a family (Ford) link with Jaguar and Aston?

Their knowledge and expertise extends across all areas of Jaguar cooling, including air conditioning, heater matrices and metal water pipes. They recommend that when the engine is out, the radiator should be checked, and replaced unless found to be excellent. You can ruin a lot of hard work by not doing so. And, of course, you should check all hoses and metal water pipes – something else that can be supplied by Autorads.

All Series XJ6 models featured brass tanks, but most other XK-engined Jaguars had pipes and nozzles which were SOFT-soldered into place. As a result, during the dismantling process, these have to be removed (in truth, they simply fall off with the effects of the blow torch) and, at a later date, replaced. This isn't a difficult job for the Autorads staff, but it does add a fair amount of time to the job. 'Our' Series 3 radiator (as seen in this section) had the various tank attachments SILVER-soldered into place, and so remain in situ during the core checking/replacement operation. Andrew Hurlock at Autorads was responsible for turning this rusty, inefficient old radiator into a beautiful and perfectly functioning unit. My thanks to him for his skill and patience.

▲ *Andrew's first task was to check the core for physical damage and rotted fins. He checked particularly at the bottom front corners, where they are vulnerable to collecting stone damage and rock salt from winter roads. Where the fins are not solid around the tubes, it allows them to overheat, which can cause splitting, as seen here. In this case the fins had rotted quite badly, meaning that the core would have to be replaced. Even if the fins are OK, a core that is more than a few years old needs checking from the end to ensure that the tubes are not blocked (see later).*

▲ *He wire-brushed the tank while applying heat from the blow torch to remove most of the paint. With the paint gone, a little more heat is applied until the original solder runs out.*

▲ *Then he took a pair of pliers to remove the tanks from the core. This is quite simple, as any few remaining globules of solder will still be quite soft. Getting the tank clean is all-important, and the job is not finished yet.*

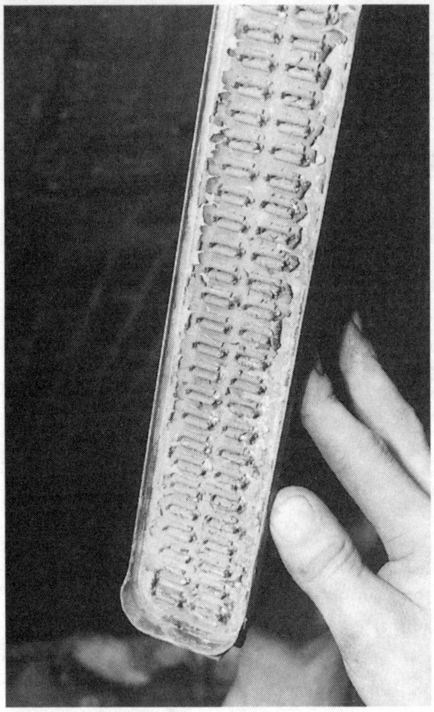

▲ *While the tanks cooled, Andrew checked the ends of the core for signs of blockage. At one end, it wasn't bad, with just two tubes being partially blocked. But at the other end, far more of the tubes were blocked and would have been contributing to any cooling problems. Any more than ten blocked tubes will cause a noticeable difference in cooling performance. Where the fins are not damaged, Autorads would usually be able to clear blocked tubes, using special rods and water under pressure, removing the necessity for core replacement. It's no use ignoring either fin damage/corrosion or blocked tubes, because it can only get worse.*

▲ *The cooled tanks were given yet another wire-brushing before being heated up again and dipped briefly into a vat of hydrochloric acid for a final clean. This effectively takes a very thin layer off the top of the brass tanks. Once it has been cooled and rinsed in the water tank, a wire brush is used to brighten the tanks and ensure that there are absolutely no traces of dirt, grime or grease.*

▲ *Each tank was given yet another going over with the wire brush. The difference is obvious between the before (at front) and after.*

▶ *Andrew checked each tank for dents, and where they were found they were tapped out gently. It's important that the 'lips' of the tanks are perfectly flat in order to make a perfect join with the new core.*

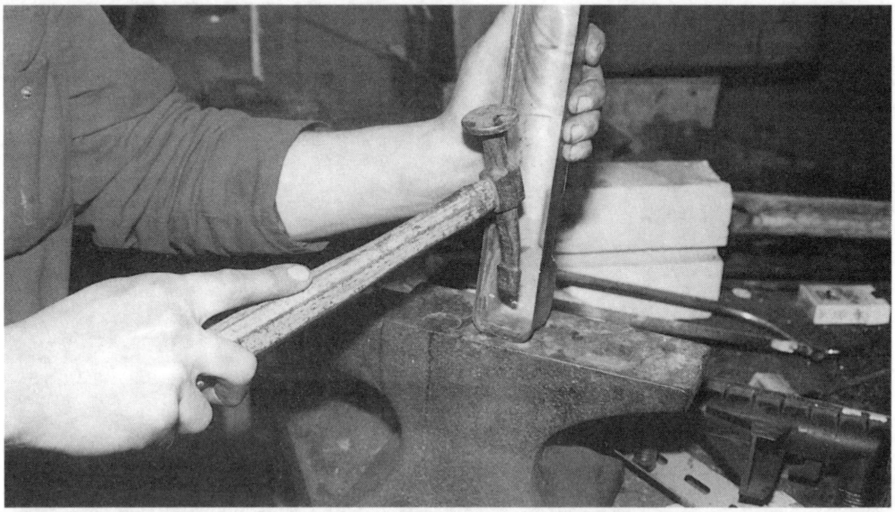

▶ *A new core was fitted. In case you can't see the difference, the old one is on the right!*

▶ *Andrew checked the tube end plates for damage which could have occurred during storage or in transit. The edges have to be perfectly straight, and if they're not, Andrew has to tap them gently out with a hammer and flat edge as shown.*

◀ *He wire-brushed the edges of the tanks to make sure that there were no foreign bodies to impede the soldering process, and then tapped the tank gently into place. It is not unknown for the tube plates to bow a little during manufacture, so he was very careful to ensure that the fit was tight all along the edge of the core.*

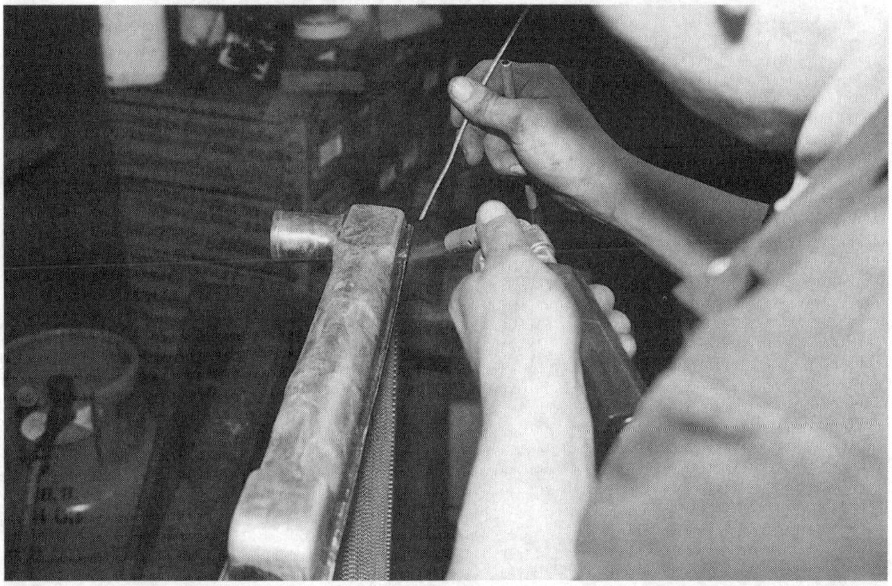

◀ *Flux was brushed on all around the tank join, and then the first of two soldered joins was made. This one actually sealed the tank, so the flame was set high and aimed deep down into the 'valley' of the tank. Andrew tinned the join as he went along, expertly using the solder and flux brush in the same hand.*

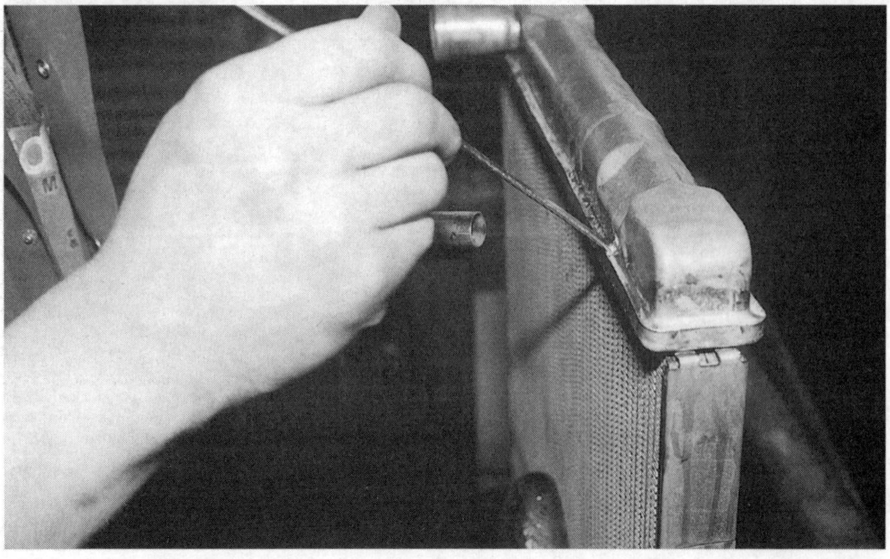

◀ *When this join had been completed, Andrew fluxed around the join again ready for the second soldered join to be made. Because this was as much for appearance as anything else, it was important not to apply too much heat and disturb the original join. As such, the flame was set low and angled high. When finished there was a nice, neat layer of solder filling the valley around the lip of the tank. Note how the top of the second join is exactly flush with the rim of the tank. It looks easy, but that's only because Andrew's had years of experience!*

◀ *Once the tank had cooled, he lightly used a dual-action pneumatic sander to remove any slight imperfections and blobs of solder from the tank, to provide a smooth surface for the paint.*

▲ *Andrew then checked the front of the radiator for any fins which may have been damaged in transit or during the fitting process. It may seem like nit-picking, but bent-over fins can soon lead to cooling problems. With the other tank soldered into place in the same professional manner, it was time to head for the acid – or rather, water – test. All orifices were plugged and a compressed air line connected to feed air into the radiator at around 20psi – in the car it usually only runs at around 10psi. The radiator was then completely immersed in water and Andrew checked carefully for signs of those tell-tale bubbles which would mean a leak. Both sides were checked and, as the glassy smooth water surface proves, there were no leaks.*

▲ *The finished article? Not quite – it still requires a coat of paint, so it first had to spend 30 minutes in the drier to get rid of every last trace of moisture.*

◀ *Once totally dry, Andrew sprayed it matt black, making four passes per side in an effort to get paint right down into all the fins. Most owners opt for this finish, although gloss black is an option.*

▲ *All Autorads radiators carry a brass plate which shows their name, telephone number and a serial number which they keep on file. This allows them to cross-reference the vehicle, date of manufacture and even the name of the employee who built the radiator – in this case …*

◀ *… Andrew, who presents the finished package here.*

COOLING TROUBLE-SHOOTING GUIDE

Symptom	*Most likely cause*	*Answer*
Overheating	Thermostat broken	Replace the thermostat
	Radiator grille blocked	Physically clear the grille
	Water pump broken	Replace the water pump
	Incorrect anti-freeze mix	Drain and top up with 50/50 solution
	Engine out of tune	Adjust plugs/points/timing etc
	Dirty engine oil	Drain and refill with new filter
	Fan belt loose	Tighten to correct tension
	Viscous fan not working	Replace fan
Overcooling	Thermostat broken	Replace the thermostat
Loss of coolant	Damaged radiator	Repair or replace radiator
	Leaking/damaged hoses	Check and replace
	Cylinder head leak	Remove cylinder head to check
	Loose head bolts	Tighten bolts in correct sequence

Chapter 15

Carburation

SU CARBURETTORS

With the exception of the 4.2-litre Series 3 XJ6 models (which featured Lucas-Bosch L-Jetronic airflow fuel injection), the pre 240 2.4-litre models (the Mk 1 and Mk 2) which were fitted with Solex carbs and some USA models which used Strombergs, all XK engines were fed their diet of fuel and air by means of SU carburettors.

The SU (Skinner Union) carburettors have an enviable reputation for being easy to maintain and easy to tune. Although the original company is not now in existence, the Burlen Fuel Company has taken over where it left off and offers a complete SU service which includes the supply of carburettors, all spares and all tuning items. Better still, there is a comprehensive technical advice service. My thanks to Burlen for their help with this section.

SPARES NOTE:
Gasket and overhaul packs are available for most SU carbs from Burlen Fuel Systems as follows:

Carburettor	*Gasket pack*	*Overhaul pack*
H6	AUE801S	AUE850
HD6	AUE805S	AUE855
HD8	AUE806S	AUE857
HS6	AUE812S	N/A
HS8	AUE813S	N/A
HIF (all types)	AUE821S	N/A

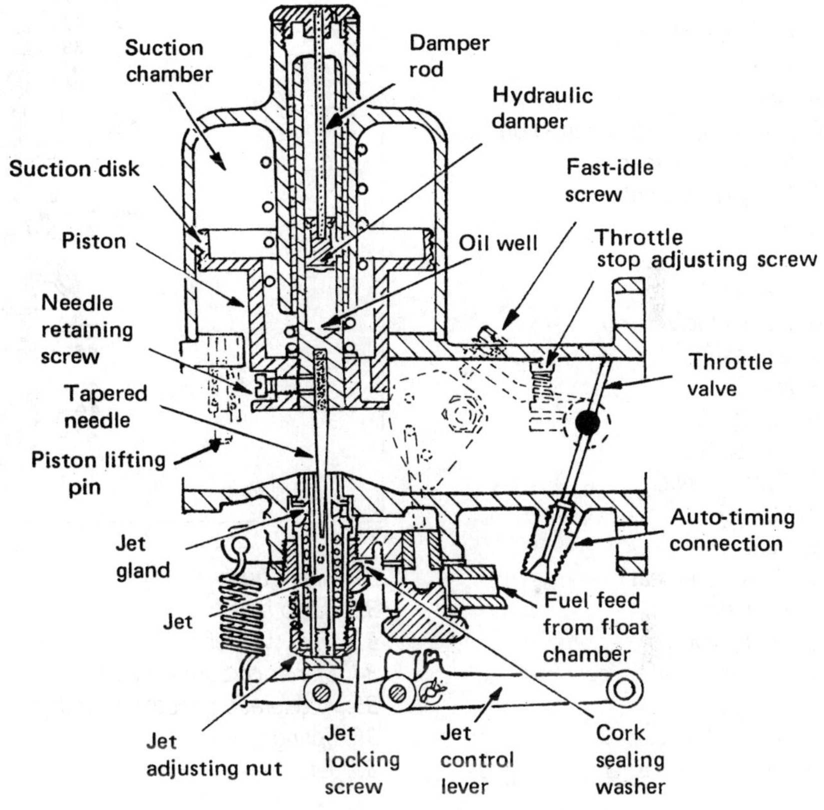

▲ The H6 SU 1¾in carburettor was fitted to all XK 120 and XK 140 models with the exception of the XK 140 fitted with the optional C-type head. This is a cross-section through an 'H' (for horizontal) carburettor. Various versions were available, including the H8 which was fitted to the XK 140 when the C-type head was specified.

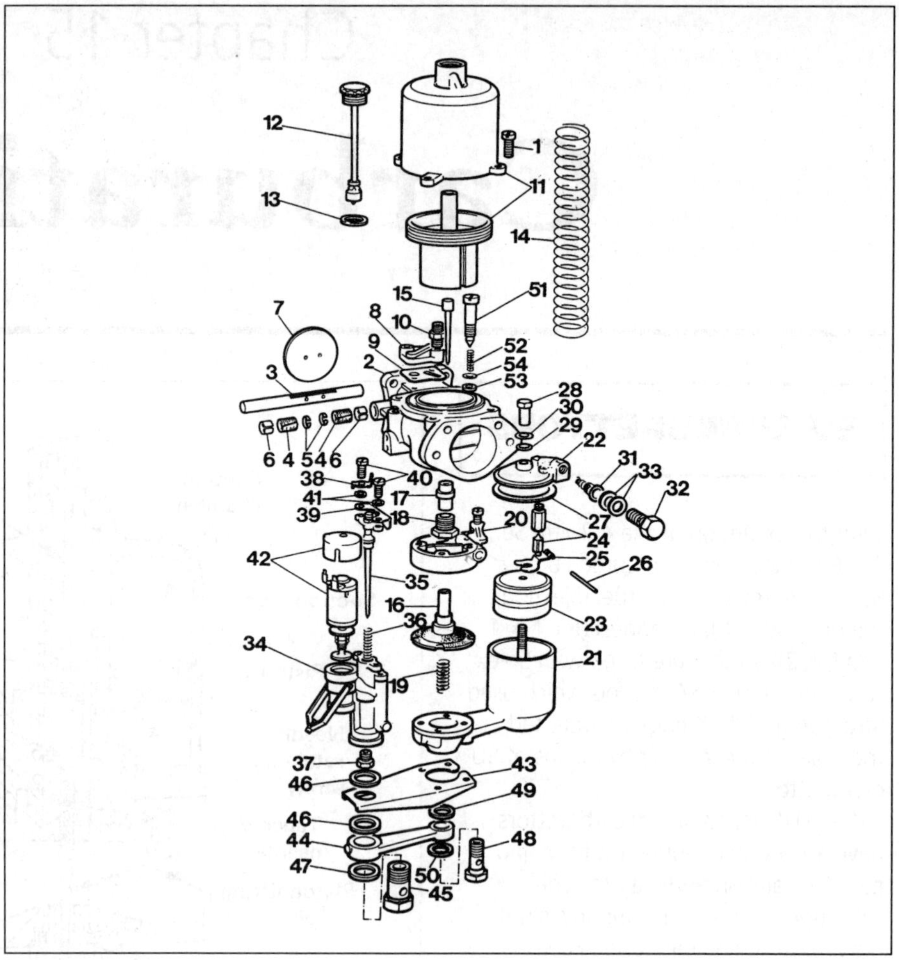

▶ *The HD carburettor ('D' for diaphragm jet) followed on and was only available in 1½in, 1¾in and 2in sizes. This diagram shows an exploded view of the SU HD8 carburettor as fitted to the XK 150S, all 6-cylinder E-types, all Mk X models, all 420 models and 2.8-litre and 4.2-litre XJ6 models up to 1975.*

1. *Screw*
2. *Body*
3. *Throttle spindle*
4. *Bush*
5. *Circlip*
6. *Circlip*
7. *Butterfly valve plate*
8. *Adaptor*
9. *Gasket*
10. *Dist vacuum pipe connection*
11. *Suction chamber*
12. *Damper assembly*
13. *Washer*
14. *Spring*
15. *Jet needle*
16. *Jet assembly*
17. *Jet bearing*
18. *Nut*
19. *Spring*
20. *Jet housing*
21. *Float chamber*
22. *Lid*
23. *Float*
24. *Needle and seat*
25. *Float arm*
26. *Float pivot pin*
27. *Gasket*
28. *Cap nut*
29. *Serrated washer*
30. *Aluminium washer*
31. *Filter gauze*
32. *Banjo bolt*
33. *Fibre washer*
34. *Starter carburettor body*
35. *Acceleration needle assembly*
36. *Spring*
37. *Jet*
38. *Spring plate*
39. *Dust shield*
40. *Screw*
41. *Shakeproof washer*
42. *Solenoid*
43. *Bracket*
44. *Connecting arm*
45. *Banjo bolt*
46. *Washer*
47. *Washer*
48. *Banjo bolt*
49. *Fibre washer*
50. *Aluminium washer*
51. *Valve*
52. *Spring*
53. *Gland washer*
54. *Dished washer*

◀ *And here's the real thing, a triple whammy of tasty carburation as fitted to the manifold of an XK 150S (and with a manual choke, the E-type).* (Courtesy Rob Beere Racing Services)

▶ *This is the SU HD6 carburettor, as fitted to the Mk 1/Mk 2 3.4/3.8-litre and 340 models and the 3.4/3.8-litre XK 150 models. This carburettor is visually very similar to the HS6 version, the main difference in construction being that the HD6 has a fixed jet and the mixture is controlled by an external screw and lever, whereas in the HS6, the position of the jet in relation to the needle can be adjusted by means of an adjusting nut at the base of the carburettor. This movement of the jet controls the strength of the mixture.*

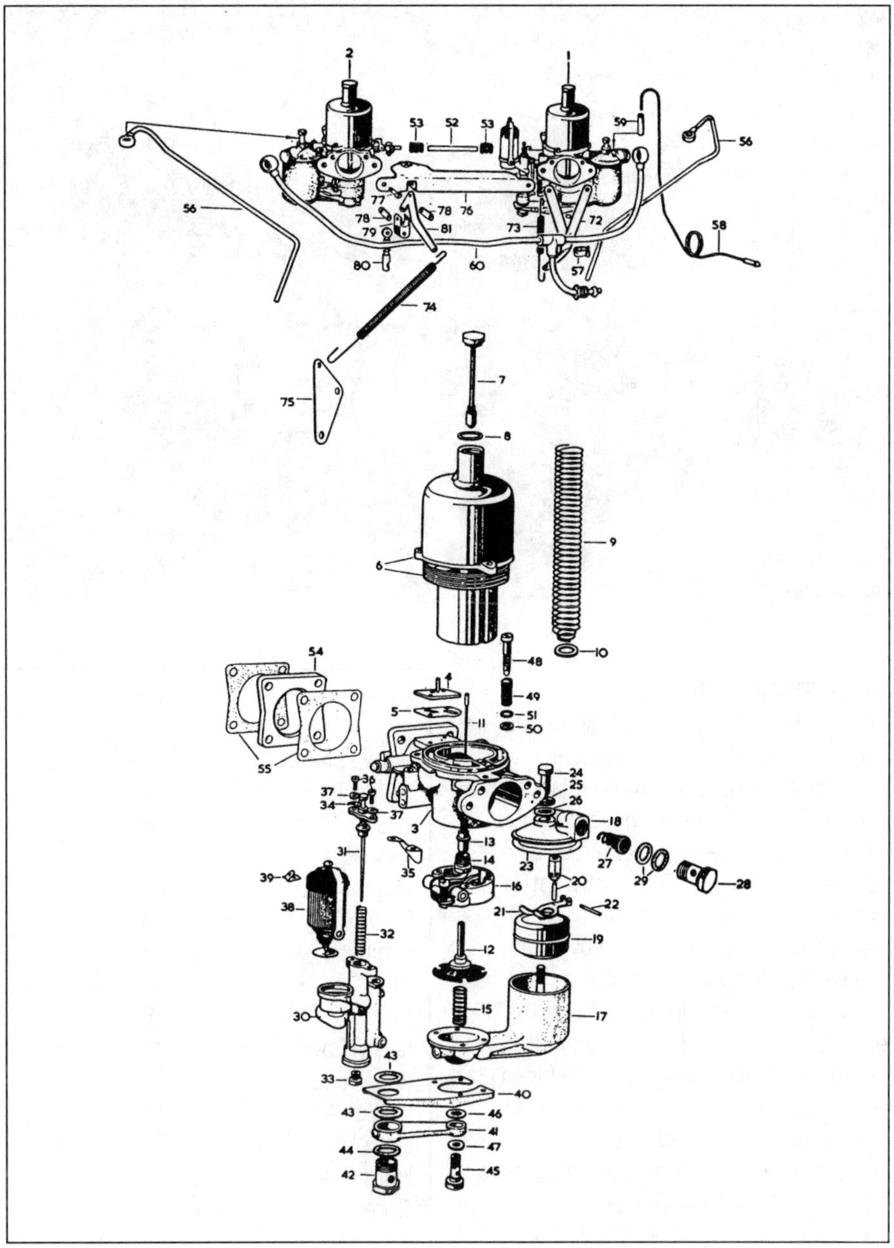

1. Front carburettor
2. Rear carburettor
3. Carburettor body
4. Ignition union adaptor
5. Gasket
6. Suction chamber and piston
7. Damper
8. Washer
9. Spring
10. Skidwasher
11. Jet needle
12. Jet
13. Jet bearing
14. Nut – jet bearing
15. Spring
16. Jet unit housing
17. Float chamber
18. Float chamber cover
19. Float
20. Float needle and seat
21. Float needle lever
22. Knurled pin
23. Gasket
24. Cap nut
25. Fibre serrated washer
26. Aluminium washer
27. Filter
28. Banjo bolt
29. Fibre washer
30. Auxiliary starting carburettor body
31. Auxiliary starting carburettor needle
32. Spring
33. Jet
34. Spring clip
35. Dust shield
36. Screw
37. Double coil spring washer
38. Solenoid
39. Spring clip
40. Bracket
41. Connecting arm
42. Banjo bolt
43. Fibre washer
44. Fibre washer
45. Banjo bolt
46. Fibre washer
47. Aluminium washer
48. Slow running control valve
49. Spring
50. Neoprene washer
51. Brass washer
52. Connecting rod
53. Connecting rod coupling
54. Manifold insulator
55. Gasket
56. Overflow pipe
57. Overflow pipe clip
58. Distributor vacuum suction pipe
59. Neoprene coupling tube
60. Petrol feed pipe
72. Front carburettor spring bracket
73. Front carburettor throttle spring
74. Throttle return spring
75. Return spring bracket
76. Throttle stop bracket
77. Dowel bolt
78. Link
79. Trunnion
80. Link rod

◀ *This photo shows a typical installation of HD6 carbs on a Mk 2, 3.8-litre engine.* (Courtesy Rob Beere Racing Services)

SU PISTON SPRINGS

The piston spring is lightly loaded and, as such, its performance is unlikely to deteriorate. Problems usually relate to mechanical damage or, in cases where the carburettors are particularly old, corrosion. Check visually for the latter and for mechanical damage (typically flattening of individual coils) by rolling the spring along a known flat surface.

The springs are colour-coded on the end coil to denote their performance and length. The table alongside, produced courtesy of Burlen Fuel Systems, shows the various details and colours. The performance of a spring can be checked by applying a reference load and measuring the spring length. However, given that all springs are readily available and that they are inexpensive, any carburettor overhaul should include new springs.

Paint colour	*Load (oz)*	*Length (in)*	*Part No.*
Black/green	5¼	2.500	AUC 5028
Light blue	2½	2.625	AUC 4587
Red	4½	2.635	AUC 4387
Yellow	8	2.750	AUC 1167
Green	12	3.000	AUC 1170
Brown	14	3.000	AUC 1168
White	18	3.562	AUC 1166
Red	4½	1.530	AUC 4355*
Yellow	8	1.530	AUC 4398*
Red/yellow	24	4.812	AUC 4478
Red/white	40½	3.875	AUC 4869
Red/green	11¼	3.875	AUC 4826
Light blue/black	4½	3.875	AUC 2107**
Light blue/red	18	3.875	AUC 4818***
Green	1	1.530	JZX 1088

* Ball-bearing type suction chamber.
** Used in place of part number AUC 5014, red/black.
*** Used in place of part number AUC 2091, blue/green.

▶ *SU introduced the 'S' version of their successful carburettor as a development of the 'H' type. It didn't replace the HD series, but ran alongside it. The main aim was to eliminate the troublesome gland seals of the previous carburettor – it can be recognised by this feature along with the different design of float chamber. This was fitted to some 3.4-litre XJ6 cars and the 4.2-litre models in 1975 prior to the introduction of the HIF7 model (and ultimately, fuel injection).*

1. *Damper*
2. *Suction chamber*
3. *Spring*
4. *Piston*
5. *Body*
6. *Float chamber lid*
7. *Float*
8. *Needle valve seating*
9. *Needle*
10. *Float chamber*
11. *Throttle valve spindle*
12. *Throttle valve*
13. *Spindle nut*
14. *Jet assembly*
15. *Flexible feed pipe*
16. *Jet fork*

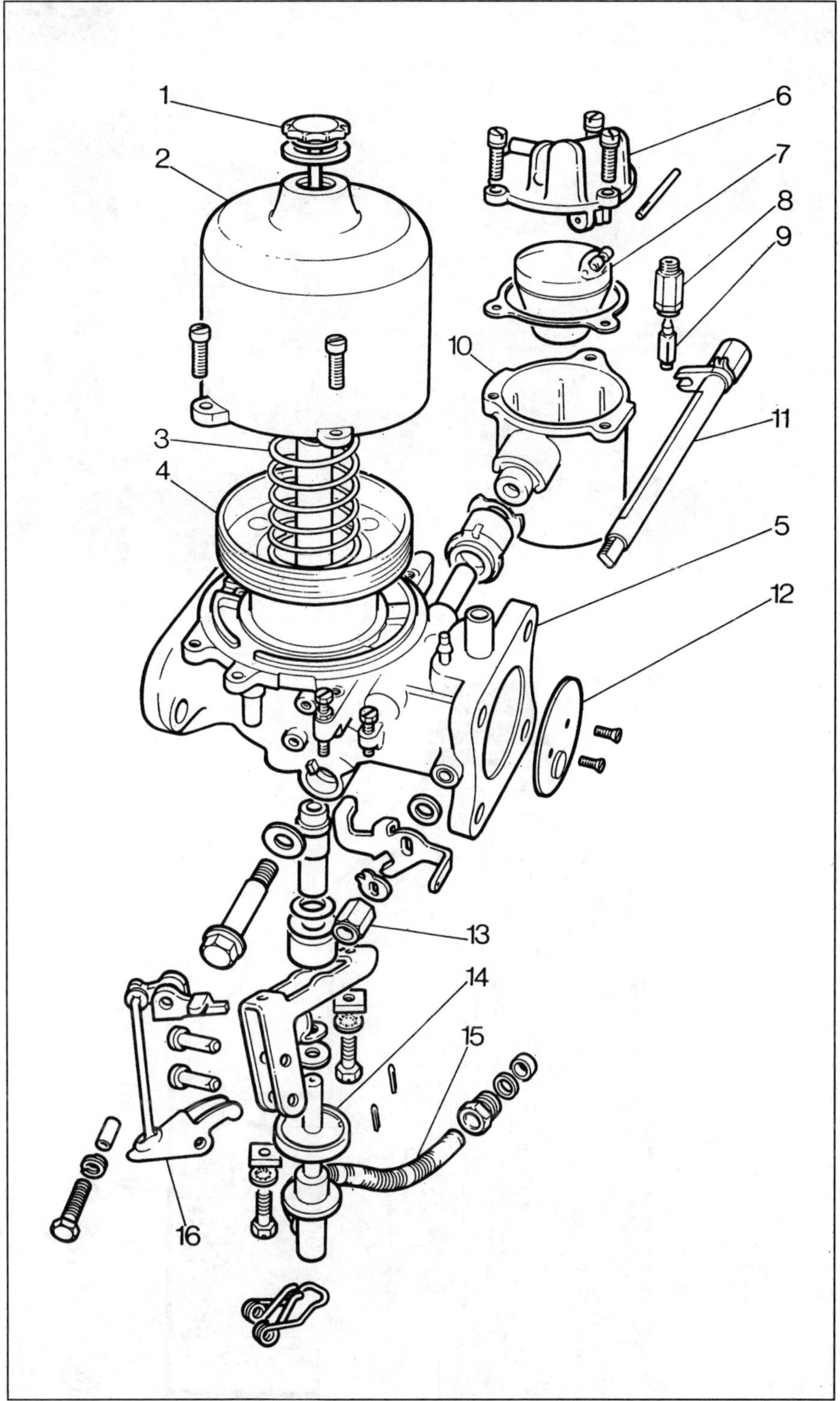

◀ *This is a pair of SU HS8 carburettors seen from the business end having just been removed from the head. Note the black plastic box at the rear centre, which is the AED (automatic enrichment device – see later). Compared to many carburettors (including Weber) the SU carbs are simple to work on, and almost all parts are still available, including complete new carbs.*

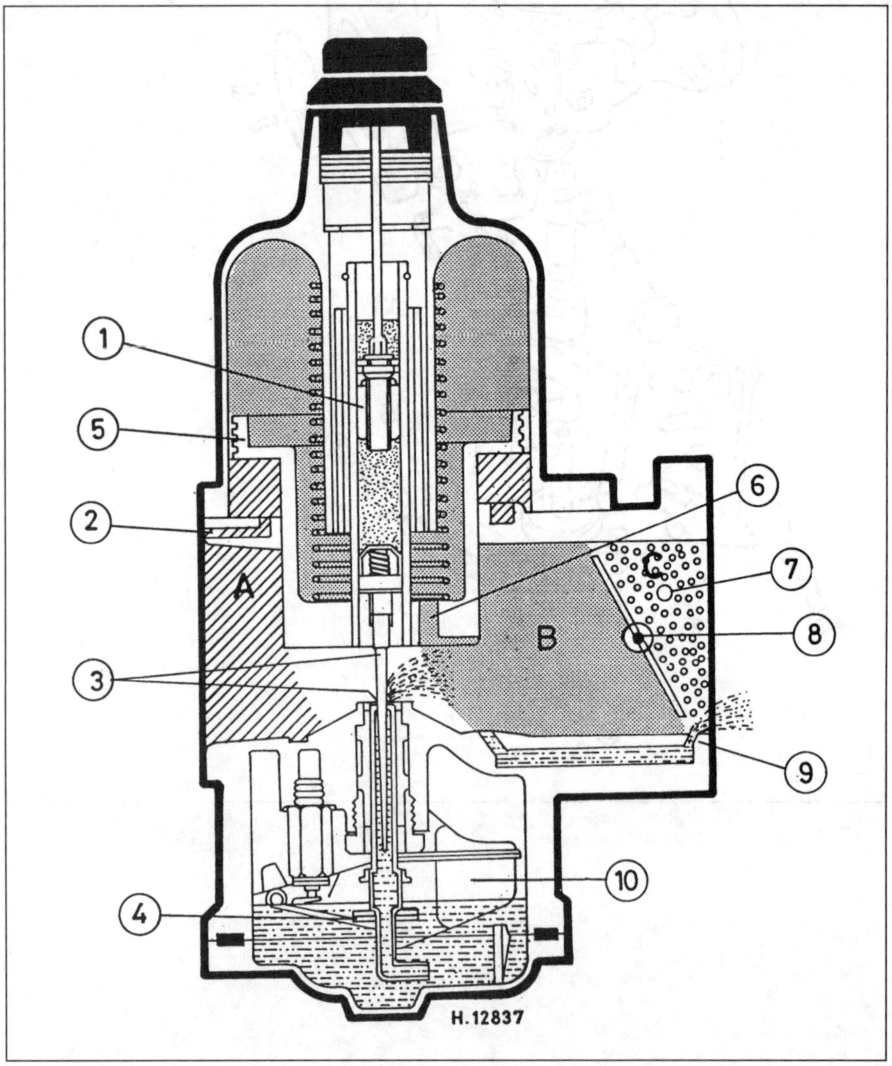

◀ *The HIF (horizontal float chamber) SU carburettors are the latest development of the SU principle. They were fitted to later 3.4-litre and 4.2-litre XJ6 models. This cross-section shows the operational parameters of the HIF carburettor.*

1. *Damper and oil reservoir*
2. *Atmosphere pressure port*
3. *Needle and jet*
4. *Bi-metal jet lever*
5. *Piston*
6. *Vacuum transfer holes*
7. *Mixture enrichment port*
8. *Throttle valve*
9. *Bypass idle port*
10. *Float*

A Atmospheric pressure
B Continuous depression
C Manifold depression

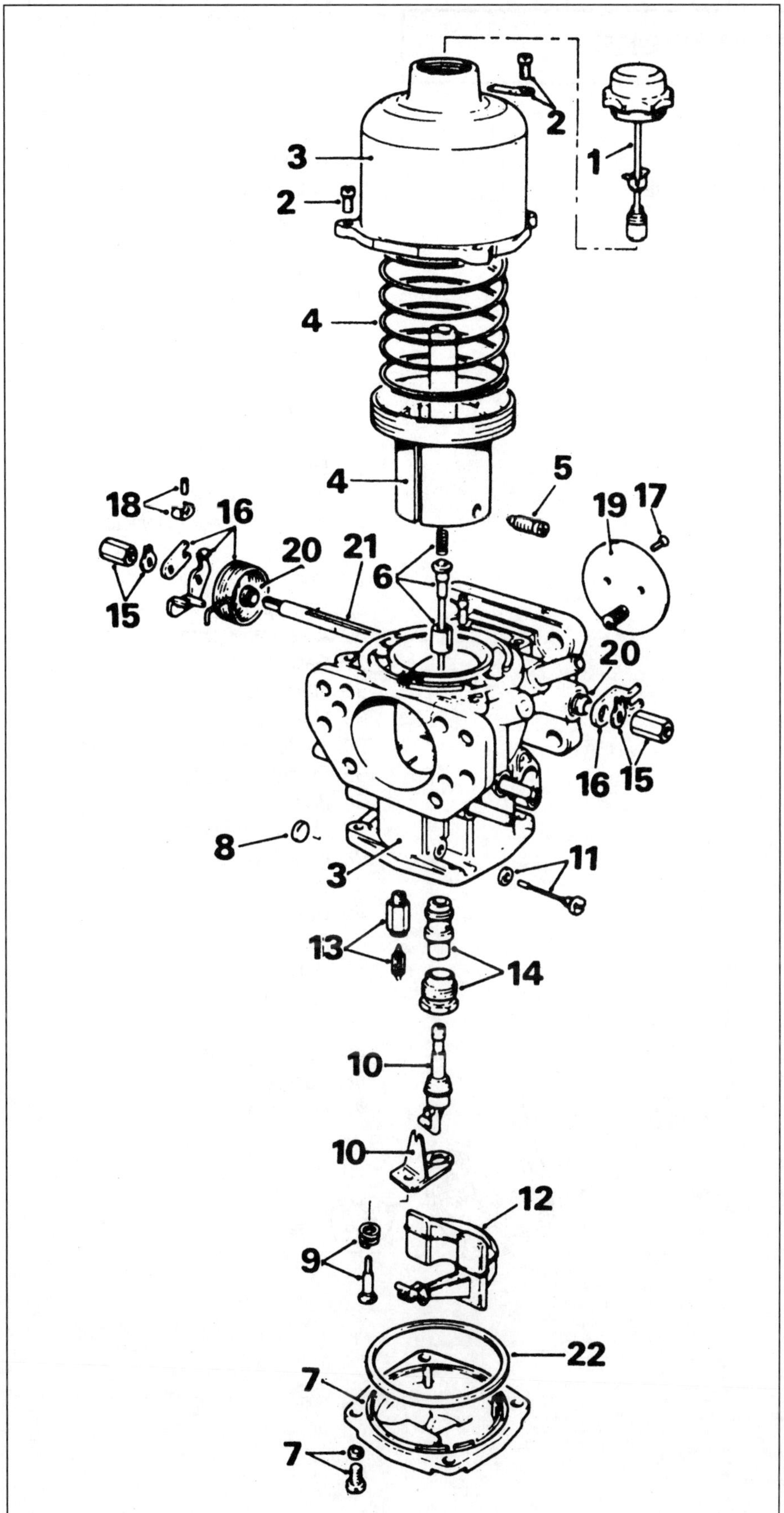

◀ *This diagram shows the complete carburettor as fitted to XJ6 models*

1. *Piston damper*
2. *Suction chamber screws*
3. *Suction chamber*
4. *Piston and spring*
5. *Needle retaining grub screw*
6. *Needle spring and guide*
7. *Bottom cover and screw*
8. *Sealing plug (jet adj screw)*
9. *Jet adjusting lever screw and spring*
10. *Jet and adjusting lever*
11. *Float pivot pin*
12. *Float*
13. *Needle valve and seat*
14. *Jet bearing and locknut*
15. *Throttle spindle nuts and lockwashers*
16. *Throttle levers and spring*
17. *Throttle disc screw*
18. *Idle speed adjustment screw and clip*
19. *Throttle disc*
20. *Throttle spindle seals*
21. *Throttle spindle*
22. *Sealing ring*

▲ *Setting up and adjusting carburettors is a tricky job, especially when you've two or three to balance up, so having the right tools for the job becomes paramount. This S-P fuel-kit includes adjustment and tune-up tools for most kinds of popular fuelling systems and includes one for Stromberg adjustment.*

SOLEX CARBURETTORS

▼ *Mk 1 and Mk 2 cars with 2.4-litre engines used Solex carburettors type B32 PBI-5 in an effort to improve fuel economy, albeit at the expense of overall performance. This diagram shows an exploded view of the Solex B32 PBI-5 – compare with any SU to see how complex it is.*

1. *Throttle chamber*
2. *Nylon insulating washer*
3. *Gasket*
4. *Control rod*
5. *Spring*
6. *Washer*
7. *Split pin*
8. *Float chamber*
9. *Non-return valve*
10. *Washer*
11. *Filter*
12. *Bolt*
13. *Main jet*
14. *Washer*
15. *Pump jet*
16. *Washer*
17. *Pilot jet*
18. *Air bleed*
19. *Starter air jet*
20. *Starter petrol jet*
21. *Washer*
22. *Choke tube*
23. *Screw*
24. *Emulsion Tube*
25. *Air correction jet*
26. *Accelerator pump injector*
27. *Gasket*
28. *Float*
29. *Float spindle*
30. *Accelerator pump*
31. *Gasket*
32. *Starter valve*
33. *Starter valve body (front carb)*
34. *Starter valve body (rear carb)*
35. *Starter valve lever*
36. *Bolt*
37. *Float chamber cover*
38. *Needle valve*
39. *Washer*
40. *Gasket*
41. *Banjo bolt*
42. *Washer*
43. *Washer*
44. *Filter*
45. *Insulating washer*
46. *Gasket*
47. *Diaphragm*

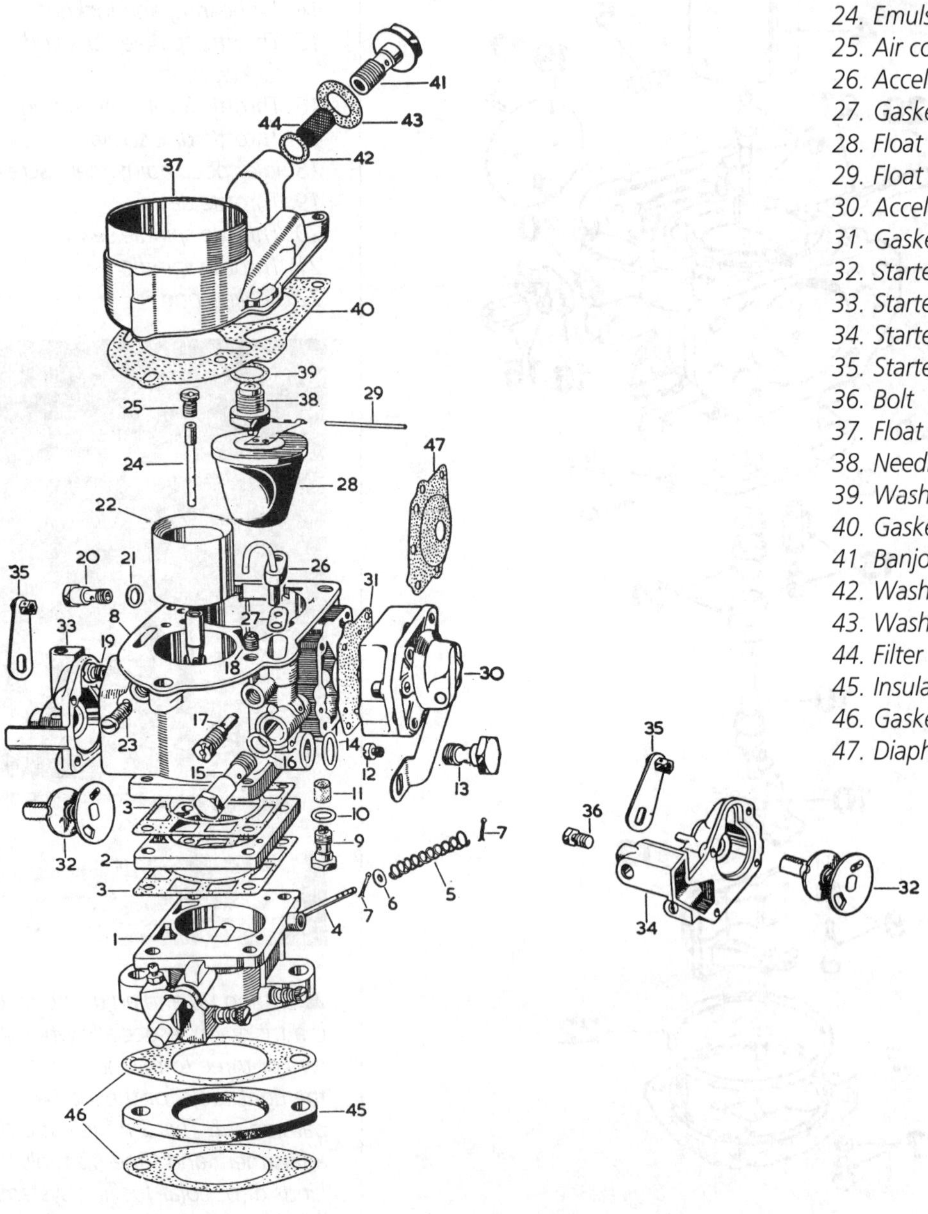

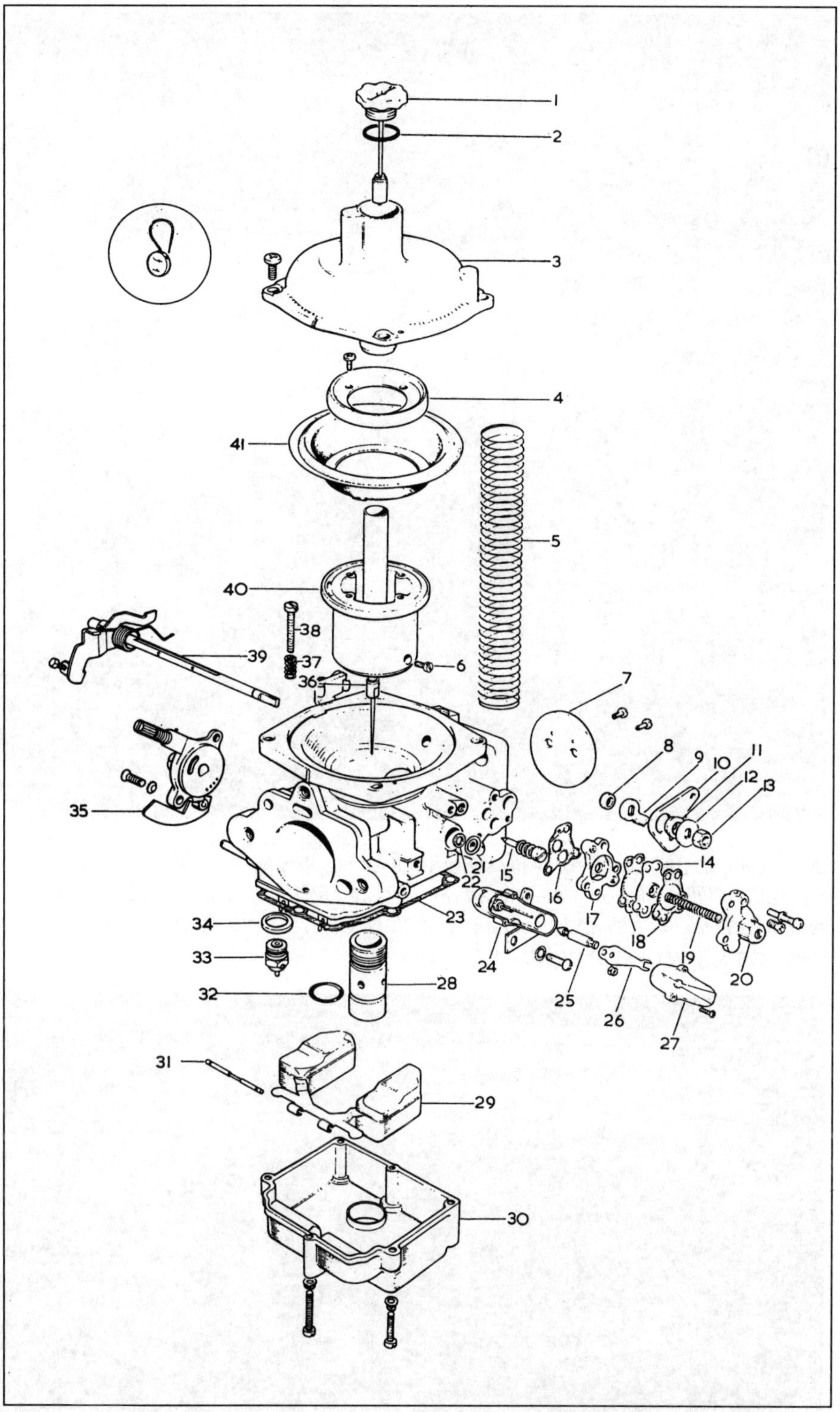

▶ *This is the Stromberg 175 CD2SE carburettor, as fitted to the E-type and XJ6.*

1. *Hydraulic damper*
2. *O-Ring*
3. *Cover*
4. *Diaphragm securing ring*
5. *Piston return spring*
6. *Needle securing screw*
7. *Butterfly*
8. *Bush*
9. *Pick-up lever*
10. *Floating lever*
11. *Washer*
12. *Shakeproof washer*
13. *Nut*
14. *Diaphragm*
15. *Idle trim screw*
16. *Gasket*
17. *By-pass valve*
18. *Gasket*
19. *Spring*
20. *Cover*
21. *Seal*
22. *Seal*
23. *Gasket*
24. *Temperature compensator housing*
25. *Tapered plug*
26. *Bi-metallic blade*
27. *Plastic cover*
28. *Jet assembly*
29. *Float assembly*
30. *Float chamber*
31. *Pivot pin*
32. *O-Ring*
33. *Needle valve*
34. *Special washer*
35. *Choke assembly*
36. *Needle*
37. *Spring*
38. *Throttle stop screw*
39. *Throttle spindle assembly*
40. *Piston*
41. *Diaphragm*

(Insert is lead seal, fitted to prevent tampering with the settings.)

▲ *As ever, in the search for really serious power outputs (with no requirement to go easy at the filling station!) Weber carburettors are favoured. This is a 3.8-litre full race engine, sucking in fuel and air via a trio of 48 DCOE Webers. And the mpg? If you have to ask, you can't afford it!*

◀ *Whilst MoT testing stations use extremely accurate equipment (costing thousands of pounds) for emissions testing, it is possible for DIY enthusiasts to check their emissions for rather less of an outlay. The Gunson's Gastester, and its digital 'Professional' sibling shown here, cost a fraction of that but enable surprisingly accurate readings to be made. Checking your emissions like this will help you tune your engine and possibly save the cost of a re-test when it comes to MoT time.*

FAULT DIAGNOSIS

Note that details of all SU carburettors, including tuning, uprating and overhauling, are to be found in the *Haynes SU Carburettor Manual* (No. 299).

'H' TYPE CARBURETTORS

Symptom	*Probable cause*	*Remedy*
Erratic running Stalling at idling Lack of power High fuel consumption	Sticking piston: Dirty piston & suction chamber Jet out of centre Bent needle	 Clean Re-centre Fit new needle
Too rich at idling Fuel leak	Jet gland leakage: Faulty top gland Dirt under top gland washer Faulty bottom gland	 Fit new gland Clean Fit new gland
Float chamber or jet flooding	Incorrect fuel level Dirty or worn float chamber needle valve Punctured float	Check and reset Clean or renew valve Fit new float

HD TYPE CARBURETTORS

Symptom	*Probable Cause*	*Remedy*
Erratic running Stalling at idling Lack of power High fuel consumption	Sticking piston: Dirty piston & suction chamber Jet out of centre Bent needle	 Clean Re-centre Fit new needle
Hesitation at pick-up	Low damper oil level Incorrect oil grade (too thin)	Top up Replace with correct grade
Float chamber or jet flooding	Dirty or worn float chamber needle valve (dirty fuel) Punctured float Incorrect fuel level	Clean or renew valve (flush system) Fit new float Check and reset

SOLEX CHOKE DETAILS

Compression ratio	*7.0:1*	*8.0:1*
Choke	23mm	24mm
Main jet	110	110
Air correction jet	200	180
Emulsion tube	14	14
Pump jet	55	55
Pilot Jet	50	50
Pilot air bleed	1.2mm	1.2mm
Needle valve	1.5mm	1.5mm
Needle valve washer	1mm	1mm
Starter petrol jet	GS.105	GS.105
Starter air jet	GA.4.5	GA.4.5

STROMBERG CARBURETTORS

Stromberg carburettors were often used by Jaguar on USA/Canada models where the emission laws meant that SU carbs were not suitable. Unfortunately, whilst the emission levels came down, so too did the engine's power and torque.

EMISSIONS

Engine exhaust emissions have become an increasingly important part of engine tuning over the years. The emissions reading is a good indicator of whether your engine is working at its best, and all the top tuners will relate how well an engine runs directly to it. Correct carburation set-up is clearly vital in making sure that the emissions are as low as possible, but it is equally important to ensure that timing, points and general engine condition (for example, making sure that the cylinder compression readings are satisfactory) are spot-on before touching the carburettors (or fuel-injection).

Moreover, emissions measurement has become part and parcel of the MoT test (*see table below*). Unless exempt (*see below*) your Jaguar's exhaust will be linked into special equipment which measures the CO (carbon monoxide) content and also the hydrocarbon content in parts per million (ppm). SU carburettors are easy to tune to get the best possible reading. Fuel-injected cars are trickier to fine tune on a DIY basis, though it should be a far more efficient system and require much less attention.

These figures represent a decrease in the previous limit for older cars from 6 per cent CO to 4.5 per cent – it doesn't sound much, but it will be difficult, if not impossible, to make a worn engine of any age achieve that figure. Later models, ie those first used after 1 October 1986, have to achieve 3.5 per cent to pass the test.

Most important, the testing procedure now reaches back to vehicles first used on or after 10 November 1973 – previously, the oldest cars tested were those used on or after 1 August 1975, a difference of almost two years. Jaguars registered before 10 November 1973, at present, will be subject only to a *visual* smoke test.

EMISSIONS REQUIREMENTS RELATING TO NON-CATALYST PETROL-ENGINED VEHICLES

Age	*Hydrocarbons*	*Carbon monoxide (CO)*
First used between 10 Nov 1973 and 1 Oct 1986	1200	4.5%
First used on or after 1 Oct 1986	1200	3.5%

CATALYST TESTING

All *new* cars registered on or after January 1992 were fitted with a catalyst exhaust system as part of EC law, and this forms a separate part of the MoT test. If you have a Jaguar fitted with a 'cat' (if it has been imported from the USA, for example), then in the UK, it will only be subject to the *normal* emissions test, NOT a catalyst-specific test.

IMPORTANT NOTE: The details here were correct at the time of going to press, but they are subject to changes in vehicle law. It is reasonable to expect that they will change at some time (indeed, they have already done so once since their introduction) and if they do, it will be to make it harder, rather than easier, to pass the test. The onus is on the owner to make sure that the engine is performing at its best at all times.

CARBURATION DETAILS

This table shows the standard SU fittings, types and sizes over the years.

Car	Year	No.	Spec	Type	Size (inches)	Needles: Rich	Needles: Standard	Needles: Weak	Piston spring
XK 120	1949/50	2		H6	1¾	–	RB	–	Red
XK 120	1951/54	2		H6	1¾	53	RF	RG	Red
XK 120 (remote air cleaner)	1951/54	2		H6	1¾	WO4	WO2	WO3	Red
XK 120 (7:1/8:1 CR) C-type head	1952	2		H8	1¾	75	VR	VE	Black/Red
XK 120 (8:1 CR) C-type head	1952	2		H6	1¾	–	RG	–	Red
XK120 (8:1 CR) (remote air cleaner)	1952	2		H6	1¾	–	DG	–	Red
XK 120 (9:1 CR) C-type	1952	2		H6	1¾	–	RC	–	Red
XK 140 (7:1/8:1 CR)	1954	2		H6	1¾	SA	SJ	LBA	Red
XK 140 (7:1/8:1 CR) C-type head	1954	2		H6	1¾	–	SR	–	Red
XK 140 (7:1/8:1 CR) C-type head, disc air cleaners	1954	2		H6	1¾	–	WO2	–	Red
XK 140 (8:1/9:1 CR) C-type head	1954	2		H8	2	75	VR	VE	Black/Red
XK 140 (7:1/8:1 CR) DHC, RHD FHC & Standard	1955	2		H6	–	–	WO2	–	Red
XK 140 (7:1/8:1 CR) LHD FHC	1955					SA	SJ	LBA	Red
XK 150 3.4	1959/62	2		HD6	1¾	WO3	TL	SJ	Red
XK 150 3.8	1960/62	2		HD6	1¾	–	TU	–	Red
XK 150S 3.8	1959/62	3		HD8	2	–	UE	–	Blue/Black
E-type 3.8	1961/64	3		HD8	2	–	UM	–	Blue/Black
E-type 4.2	1965/68	3		HD8	2	–	UM	–	Blue/Black
Mk X 3.8	1964	3	AUD 111	HD8	2	–	UM	–	Blue/Black
Mk X 3.8 (8:1/9:1 CR)	1964	3	AUD 144	HD8	2	–	UM	–	Blue/Black

CARBURATION DETAILS CONTINUED

Car	Year	No.	Spec	Type	Size (inches)	Needles Rich	Needles Standard	Needles Weak	Piston spring
Mk X 3.8 Auto & Overdrive	1964	3	AUD 156	HD8	2	–	UM	–	Blue/Black
Mk X 3.8	1964	3	AUD 157	HD8	2	–	UM	–	Blue/Black
Mk X 4.2 (8:1/9:1 CR)	1964	3	AUD 144	HD8	2	–	UM	–	Blue/Black
Mk X 4.2 Auto & Overdrive	1964	3	AUD 156	HD8	2	–	UM	–	Blue/Black
Mk X 4.2	1964	3	AUD 157	HD8	2	–	UM	–	Blue/Black
240	1967/68	2	AUD 256	HS6	1¾	–	TL	–	Red
240 Auto	1967/68	2	AUD 297	HS6	1¾	–	TL	–	Red
240	1968/69	2	AUD 309	HS6	1¾	–	TL	–	Red
240 Auto	1968/69	2	AUD 310	HS6	1¾	–	TL	–	Red
3.4	1963/64	2	AUD 109	HD6	1¾	–	TL	–	Red
3.4 S-type	1967/68	2	AUD243	HD6	1¾	–	TL	–	Red
3.8 Mk 2	1963/64	2	AUD 109	HD6	1¾	–	TL	–	Red
3.8 Mk 2 (Cooper air cleaner)	1964	2	AUD 155	HD6	1¾	–	TM	–	Red
3.8 S-type (Paper air cleaner)	1964	2	AUD 153	HD6	1¾	–	TL	–	Red
3.8 S-type (Oil bath air cleaner)	1964	2	AUD 154	HD6	1¾	–	CI	–	Red
3.8 S-type (Paper air cleaner)	1967/68	2	AUD 243	HD6	1¾	–	TL	–	Red
340 (AC Paper air cleaner)	1967/68	2	AUD 241	HD6	1¾	–	TM	–	Red
340 (8:1/9:1 CR, AC Paper air cleaner)	1967/68	2	AUD 242	HD6	1¾	–	CI	–	Red
420 Manual	1967/68	2	AUD 239	HD8	2	–	UM	–	Blue/Black
420 Auto	1967/68	2	AUD 245	HD8	2	–	UM	–	Red/Green
420G Manual	1967/68	3	AUD 157	HD8	2	–	NA	–	Blue/Black
420G Auto	1967/68	3	AUD 156	HD8	2	–	UM	–	Blue/Black
2.8 XJ6	1968/71	2	AUD 321	HD8	2	–	UVV	–	Blue/Black
2.8 XJ6	1971/72	2	AUD 415	HS8	2	–	BAU	–	Blue/Black

Car	Year	No.	Spec	Type	Size (inches)	Needles Rich	Needles Standard	Needles Weak	Piston spring
2.8 XJ6 (LHD)	1972/73	2	AUD 537	HS8	2	–	BBL	–	Red/Green
4.2 XJ6	1968/71	2	AUD 357	HD8	2	–	UM	–	Red/Green
4.2 XJ6 (LHD)	1972/73	2	AUD 538	HS8	2	–	BBK	–	Red/Green
4.2 XJ6	1971/73	2	AUD 397	HS8	2	–	BAW	–	Red/Green
4.2 XJ6	1973	2	AUD 647	HS8	2	–	BAW	–	Red/Green
4.2 XJ6	1973/	2	AUD 653	HS8	2	–	BCC	–	Red/Green

The rich, standard and weak needle references apply to the engine as it was originally supplied, not to any tuned versions. The standard needle will give performance as originally intended, whereas the rich needle will give a richer fuel/air mixture and greater power throughout the rev range. However, the weak needles are **not** economy needles. They are designed to compensate for operation at high altitudes and it is by no means certain that better fuel economy will result from their use.

1975-ONWARD

Car	Year	No.	Spec	Type	Throttle diameter	Needle	Piston spring	Jet
*XJ6 3.4	1975/76	2	AUD 710	HS8 (AED)	2in	CUD 1169	AUC 4826	CUD 2753
XJ6 3.4	1975/76	2	FZX 1001R FZX 1049R	HIF7 (AED)	1⅞in	NZX 8013	AUD 4355	LZX 1068
XJ6 3.4	1976/79	2	FZX 1053	HIF7 (AED)	1⅞in	NZX 8021	AUD 4355	LZX 1068
XJ6 3.4	1979	2	FZX 1330	HIF7	1⅞in	NZX 8021	AUD 4355	LZX 1068
XJ6 4.2	1975	2	AUD 667	HS8 (AED)	2in	CUD 1150	AUC 4826	CUD 2753
XJ6 4.2	1975/76	2	FZX 1049	HIF7 (AED)	1⅞in	NZX 8013	AUD 4355	LZX 1068
XJ6 4.2	1976	2	FZX 1252	HIF7 (AED)	1⅞in	NZX 8023	AUD 4355	LZX 1068
Daimler limousine		2	FZX 1459	HIF44	44mm	NZX 8068	AUD 4398	LZX 1028

NOTES:

* Fitted with Lucas/Bosch fuel-injection.

Where a carburettor reference number is given, there will also be a suffix, being F, C or R, standing for front, centre or rear, depending on whether it is a two or three-carburettor set-up.

The later HIF carburettor throttle diameters were classified in metric terms with the HIF44 being 44mm (1.73in).

Chapter 16

Auxiliary Starting Devices

To aid cold starting performance, the SU carburettor has used either a thermostatic starting carburettor (auxiliary enrichment device) or, latterly, an automatic enrichment device (AED). Both were designed to provide a richer mixture at low temperatures. Before adjustments are made to the auxiliary carburettor or AED, it is vital that the main carburettors are perfectly tuned and balanced.

AUXILIARY CARBURETTOR BASIC OPERATION

With the engine stationary and the solenoid coil energised, the armature is retracted into the bore of the coil against the loading of the conical spring, thus lifting the disc valve from its seat. The coil is switched on by a thermostatic switch located in the engine cooling system.

When the engine is rotated by the starter, air is drawn through the air intake and past the jet where it is mixed with fuel. The mixture flows upward, past the shank of the needle and out through the disc valve into the inlet manifold via the external feed pipe. The level of fuel at the jet is controlled by the main carburettor float chamber inlet valve and is above the jet when the engine is stationary, giving a reservoir of fuel in the well of the auxiliary carburettor. When starting with the device in operation, this fuel is drawn into the induction manifold, creating the rich mixture necessary for instant starting.

With the engine running the needle responds to variation in inlet manifold depression against the load of the compression spring. With increasing depression, the needle moves downward into the jet until the head abuts the adjustable stop. This weakens the mixture. The mixture is at its weakest at high depression (idling, with the throttle almost closed) and richest at low depression (large throttle openings). The engine is thus supplied with an additional quantity of mixture at a strength to suit its demands.

If the solenoid is energised while the engine is idling, the disc valve will not normally lift, as the force due to the high manifold depression, combined with the spring load is greater than the attraction force generated by the solenoid. Momentary opening of the throttle will reduce the depression and permit the valve to lift. It will remain lifted when the throttle is closed.

NOTE: When it comes to overhaul, there is a gasket pack available (Part No. AUE 807). Five different springs were used for auxiliary carburettors over the years, as follows:

Paint colour on end coil	*Load (oz)*	*Length (mm)*	*Part No.*
White	1.75	1	AUC 1195
Blue	2.25	1	AUC 1041
Yellow	2.73	1	AUC 5021
Red	3.25	1	AUC 3427
Green	3.75	1	AUC 3217

AUTOMATIC ENRICHMENT DEVICE (AED) BASIC OPERATION

With the engine cold, the main valve is opened by the main bi-metal unit, and the metering needle is raised by the diaphragm spring until it is restrained by the secondary bi-metal spring. When the engine is rotated on the starter, air is drawn through the air valve and main valve and into the inlet manifold. The resulting depression draws fuel from the fuel well, through the orifice and jet, up the jet tube and into the inlet manifold, producing a very rich mixture.

When the engine starts, the inlet manifold depression draws the diaphragm down and allows the tapered metering needle to move downward under its spring tension. During initial starting, the mixture is further enriched by fuel from the fuel well. However, when this is exhausted, air passes through the fuel well and small orifice. As the engine temperature increases, the main bi-metal unit gradually closes the main valve until it is completely closed at normal operating temperature. An air bleed valve through the bi-metal chamber ensures that the main valve remains closed during normal running.

Enrichment for acceleration during the warm-up stage is catered for by inlet manifold depression acting on the diaphragm and main valve. When the

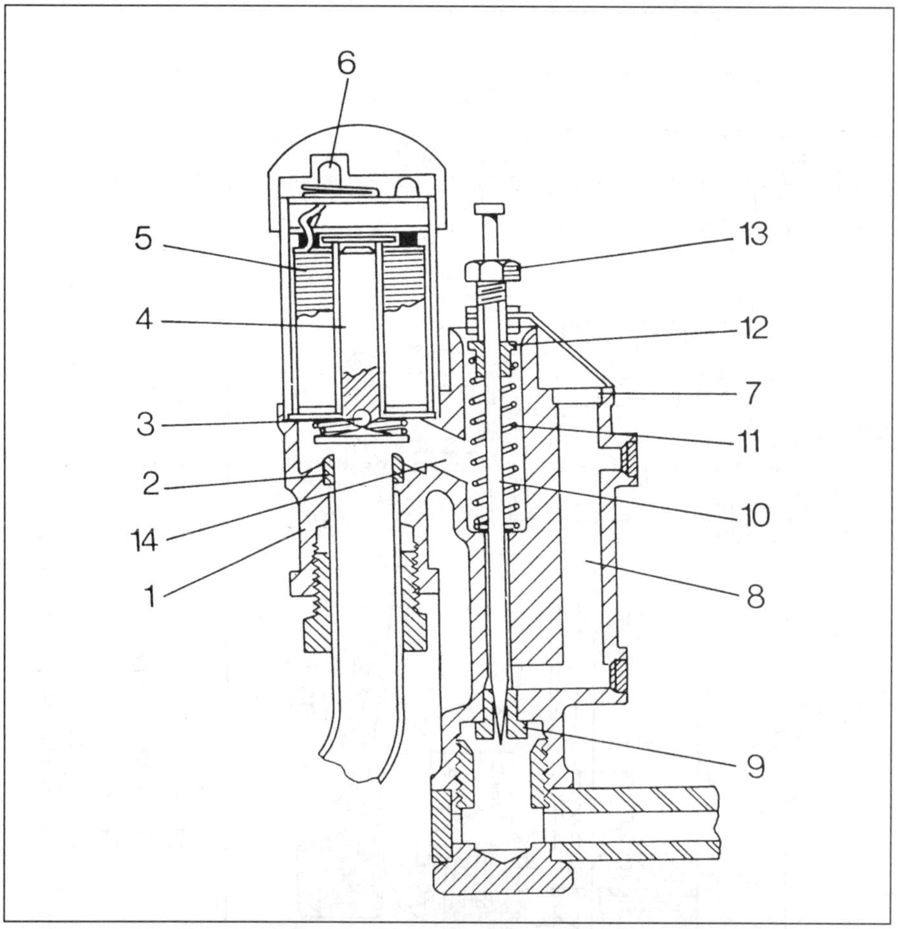

throttle is opened, the metering needle will rise and the main valve will open due to the reduced depression tending to keep it shut.

When the hot engine is stopped, heat dissipation from the AED is matched to that of the engine, so that the device will only operate at the required temperature.

◀ *The auxiliary carburettor is a relatively simple device, as can be seen in this cutaway diagram, as fitted to a HD8 carburettor.*

1. *Main body*
2. *Valve seating*
3. *Solenoid valve*
4. *Solenoid core*
5. *Solenoid winding*
6. *Solenoid terminals*
7. *Air intake*
8. *Air intake passage*
9. *Jet*
10. *Metering needle*
11. *Spring*
12. *Disc*
13. *Adjustable needle stop*
14. *Interconnecting passage*

◀ *This photo shows the carburettor from the cylinder head side, with the auxiliary carburettor seen on the left. Though its function remained the same, it tended to move about a bit, as it were; for example, on early XK 120s, it was fitted to the rearmost carburettor, but on later cars it was moved to the front carburettor.*

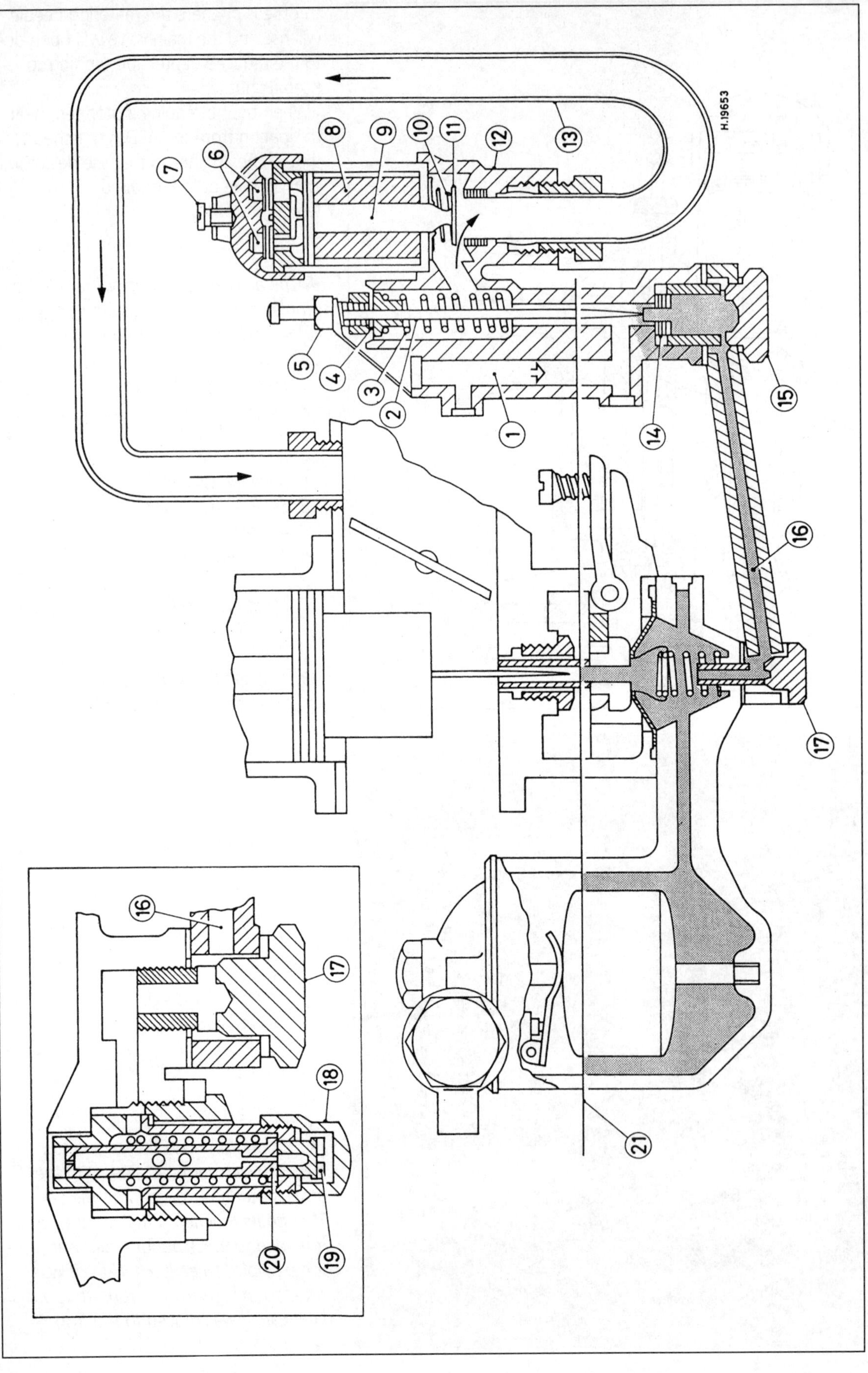

1. Air inlet
2. Tapered metering needle
3. Spring
4. Suction disc
5. Adjustable stop nut
6. Terminals
7. Clamping screw
8. Solenoid
9. Iron core plunger
10. Conical spring
11. Valve
12. Body
13. External feed pipe
14. Jet
15. Banjo bolt
16. Fuel channel in mounting arm
17. Banjo bolt
18. Cap nut
19. Jet adjusting nut
20. Jet flange
21. Fuel level

This photo shows an AED (the black plastic box in the centre) fitted to a pair of SU HS8 carbs.

AED PROBLEM SOLVER

Symptom	***Probable cause***	***Suggested remedy***
A. Engine fails to fire when cold.	1. Inadequate or no fuel supply to the AED.	1. Slacken the filter plug on the AED. Crank the engine; fuel should leak from the plug. If no fuel, check the system leading to the AED. If fuel is present, tighten the plug. Then carry out check 2 under symptom A. 2. Note: This check is not applicable when AED temperature is above 35°C (95°F). Crank the engine for several seconds. Remove the pipe at the inlet manifold. If fuel is present, it indicates that the AED is satisfactory and the cause for failure to start must be traced to some other source. If no fuel is present this indicates a faulty AED. Proceed as follows: Remove the float-chamber lid and check that the needle valve and float are free to move. If there is no fuel in the float chamber, check that the fuel filter is clean and that ample fuel is being delivered to the unit. If adequate fuel is being delivered to the float chamber this indicates a malfunction elsewhere in the unit (provided the temperature is within the unit operating range) and the unit should be replaced.
B. Engine fires but fails to keep running when cold.	1. Sticking or faulty needle valve or float.	1 Check the following: a) Remove the vent pipe from the float chamber and check that fuel is not discharged when cranking the engine. b) If fuel is discharged, check the operation of the needle valve and float as detailed under A item 2.
	2. Faulty air flap valve.	2. Remove the air inlet elbow on the AED and check that the flap valve is free to move and return under spring load.
	3. Inadequate fuel supply.	3. Check for lack of fuel as detailed under A.

Symptom	Probable cause	Suggested remedy
C. Engine fails to start when hot or part warm, or fires and fails to keep running.	1. Incorrect starting procedure.	1. Crank the engine and open the throttle slightly. If the unit is badly over-choked open the throttle fully. Do not over-rev. If the engine starts but fails to keep running, carry out check 4 below.
	2. Leaks from pipe layout.	2. Ensure that all pipes are correctly fitted, particularly the pipe between the hot air pick-up and the AED. Rectify as necessary, ensuring air-tight joints. Note: Air leaks at the hot air entry will result in excessive fuel consumption and the possibility of stalling at certain engine temperatures.
	3. Before carrying out further checks ensure that the AED is not the cause of failure to start.	3. Remove the pipe from the AED to inlet manifold. Blank off the aperture in the manifold. If the engine starts this indicates that the AED is at fault, and further checks must be made. However, if the engine is flooded, it may be necessary to crank for several seconds with a slightly open throttle before the engine will fire.
	4. Sticking or faulty needle valve or float.	4. Check the following: a) Remove the vent pipe from the float chamber and check that fuel is not discharged when cranking the engine. b) If fuel is discharged, check the operation of the needle valve and float. Remove the float chamber lid. Check that needle valve and float are free to move. If there is no fuel in the float chamber, check that the fuel filter is clean and that ample fuel is being delivered to the unit. See the servicing instructions.
	5. Faulty needle diaphragm.	5. Carry out checks as detailed under B item 4.

▶ *This diagram shows the component parts of the AED.*

1. *Heat insulation cover*
2. *Blanking plugs*
3. *Top Cover*
4. *Secondary bi-metal adjusting screw*
5. *Main bi-metal adjusting screw*
6. *Gasket*
7. *Long screw*
8. *Short screw*
9. *Spring washer*
10. *Main bi-metal unit*
11. *Spring seat*
12. *Loading spring*
13. *Needle assembly*
14. *Valve body*
15. *Gasket*
16. *Screw*
17. *Spring washer*
18. *Plain washer*
19. *Main valve and diaphragm*
20. *Diaphragm clamp ring*
21. *Air entry pipe*
22. *Spindle*
23. *Return spring*
24. *Hot air inlet stub*
25. *Clamp*
26. *Screw*
27. *Needle diaphragm*
28. *Spring*
29. *Cap*
30. *Locating dowel*
31. *Screw*
32. *Spring washer*
33. *Float chamber cover*
34. *Nylon filter*
35. *Plug*
36. *Washer*
37. *Screw*
38. *Spring washer*
39. *Float*
40. *Float needle*
41. *Gasket*
42. *Main body*

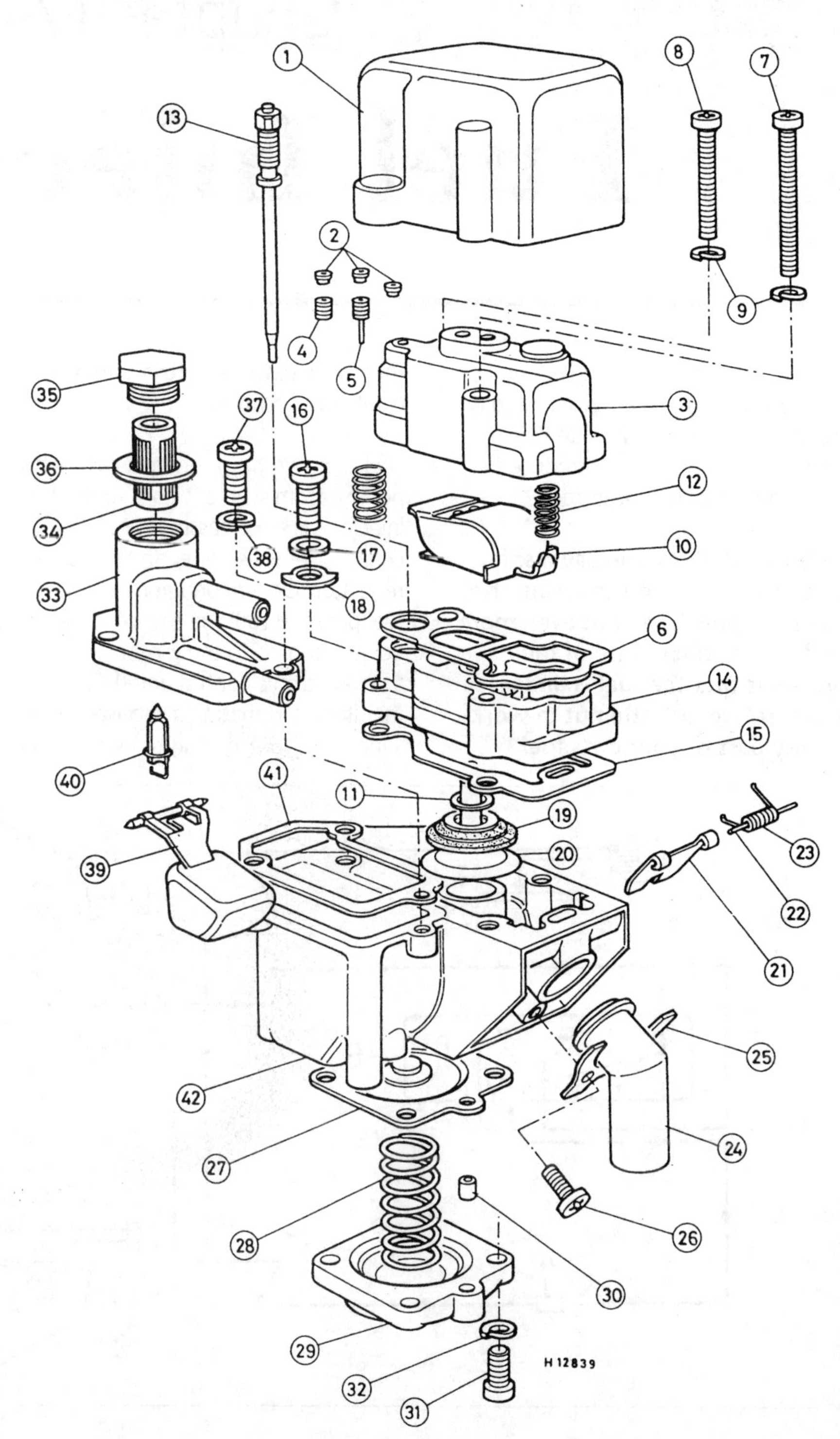

Chapter 17

Fuel Injection

Lucas-Bosch L-Jetronic airflow fuel injection was only fitted as standard in the UK to late model 4.2-litre XJ6 models, although it was introduced earlier on North American cars.

SAFETY NOTE: All the usual safety precautions relating to carburettor engines apply here, but even more so, because the nature of the system is that the fuel is under pressure. Do not attempt to work on any part of your car's fuel injection system unless you are qualified to do so.

Fuel is drawn from the tanks by electric pumps, from where it goes to the fuel injector rail after passing through a changeover valve, filter and pressure regulator. On air-conditioned cars, the fuel passes through a fuel cooler before being returned to the tank.

An electronic control unit (ECU) monitors information it receives from various sensors relating to engine load, speed, coolant and induction air temperatures and throttle opening, and from this it allows more or less fuel to be supplied to the cylinders, as required.

There are six electro-mechanical fuel injectors (one per cylinder) which form part of a fuel rail and a cold-start injector, acting as a choke would on a carburettor.

TECH SPEC

Idle speed	750 rpm
CO level	1.25–1.75%
Fuel pressure	2.55 bar (+/- 0.05 bar)

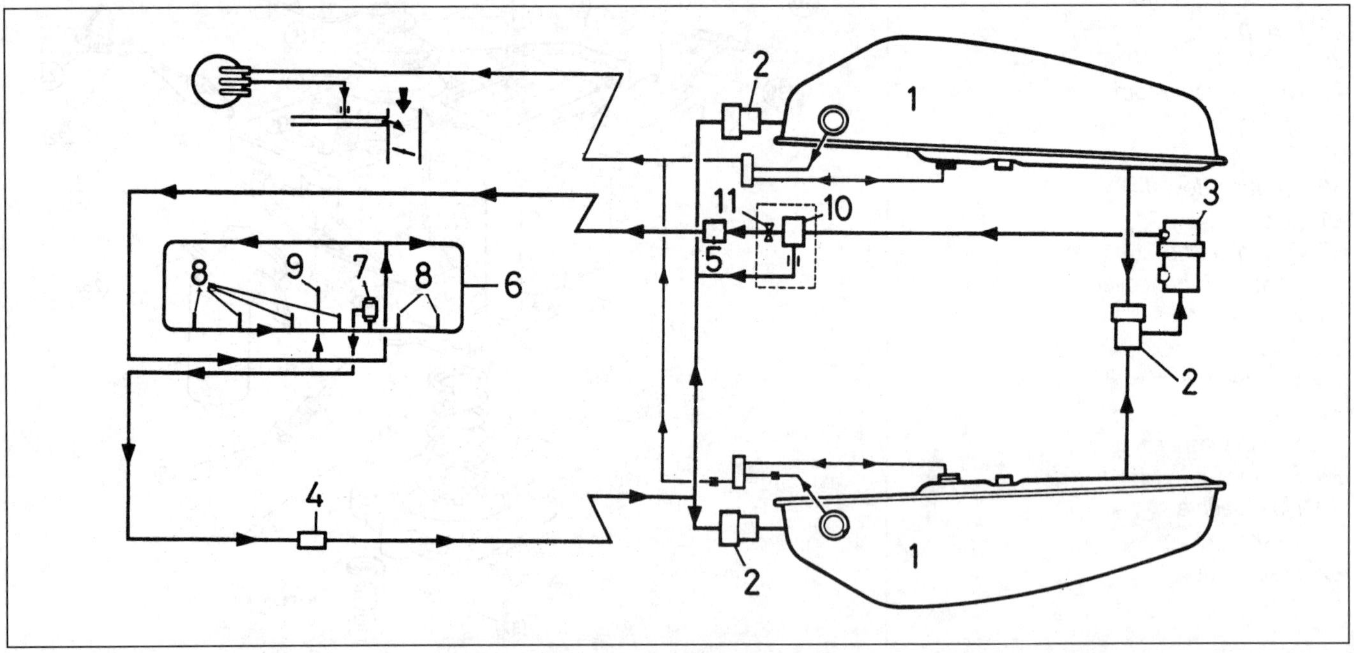

▲ *Though fuel injection systems are inherently more complex than comparable carburettor set-ups, they have several advantages – most notably better, smoother performance, better mpg and low emissions levels. The Lucas system as fitted to the XJ6 is far from being the most complex system. In most cases, the cause of problems can be traced to faulty sensors rather than a faulty electronic control unit (ECU), despite the latter being what many owners worry about most. This diagram shows the sensors used on the XJ6.*

1. *Thermo time switch*
2. *Fuel injectors (6)*
3. *Auxiliary air valve*
4. *Air-flow meters*
5. *Overrun valve*
6. *Cold start injector*
7. *Throttle switch*
8. *Full load vacuum switch*
9. *Fuel pressure regulator*
10. *Water temperature sender.*

DEPRESSURISING THE INJECTION SYSTEM

Early models did not feature a separate fuel pump relay, so the best way to disable the pump is to access it in the spare wheel well and remove the earth lead. On later models, remove the fuel pump relay from the bulkhead in the engine bay. In both cases, with the fuel pump disabled, crank the engine for 5/10 seconds. Replace the earth lead/relay, and at this point the system should be depressurised and will remain so until the next time the engine is cranked. Though the full pressure will not be present, there will inevitably be some spillage from various components if they are disconnected/removed. Therefore, it is vital to take all the usual safety precautions.

FUEL INJECTION BASIC COMPONENTS

ELECTRONIC CONTROL UNIT (ECU)

The ECU is a very complex box of electronic components and is non-serviceable by the DIYer. However, most fuel injection problems tend to relate more to the sensors which supply information to the ECU than that device itself. So, it's usually wise to assume that a fuel injection fault is being caused by a (relatively) cheap sensor rather than the expensive ECU.

FUEL RETURN VALVES

There are two fuel return valves, one situated under each rear wheel arch. It is important to note that they are 'handed' and, as such, are not interchangeable. This should also be mentioned when ordering replacement valves.

THROTTLE SWITCH

The basic aim of a throttle switch is to inform the ECU when the throttle is fully open so that maximum fuel enrichment is provided. European specification cars actually featured two throttle switches, the extra one of which was linked to the manifold vacuum. Subject to an adequate engine speed being maintained, the fuel is shut-off totally on the overrun.

INERTIA SWITCH

As a safety measure in case of an accident, the fuel injected XJ6 also features an inertia switch. This is fitted in the passenger side of the vehicle near the door pillar. Its object is to cut the power to the fuel pump in an emergency. Access for resetting can be gained simply by removing its cover.

AUXILIARY AIR VALVE

This is an idle air valve which allows extra air past the throttle butterfly to compensate for the additional fuel injected at very low temperatures. It is solenoid operated and controlled by the ECU. The air allowed to enter the engine via this valve is measured by the ECU via the airflow meter.

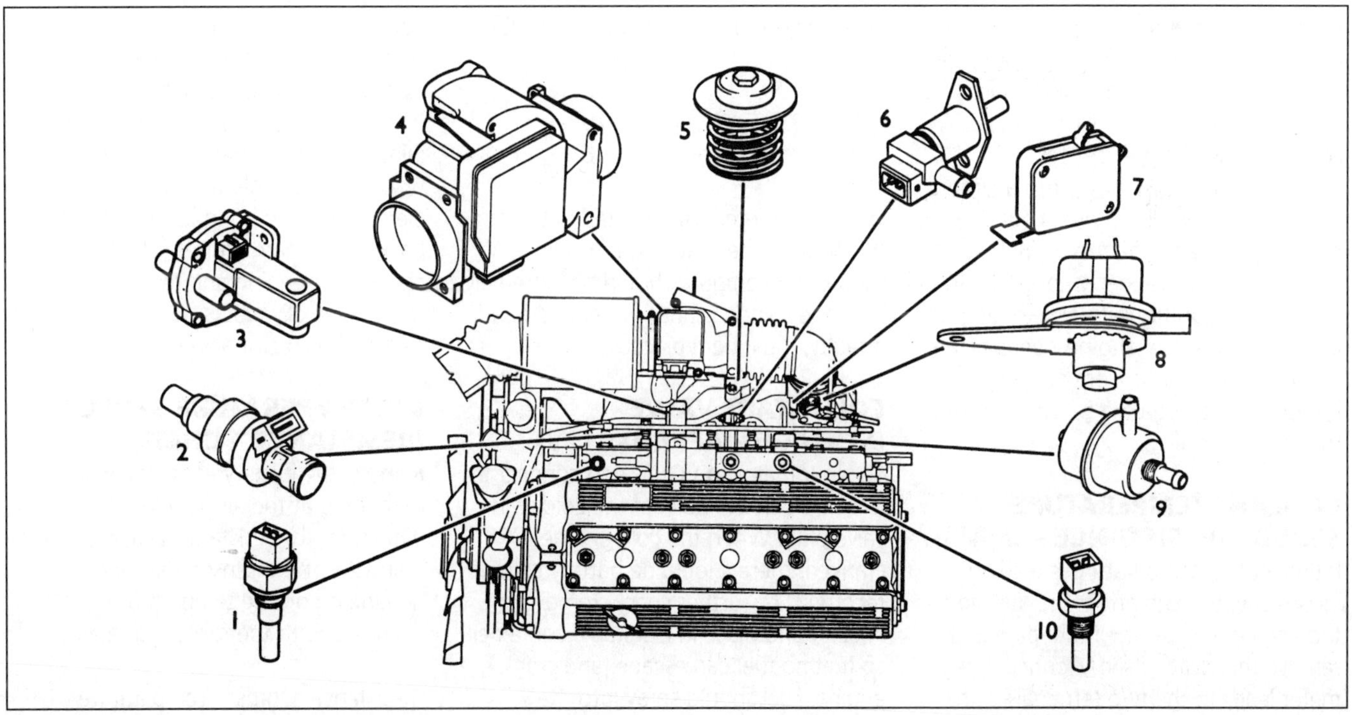

▲ *Schematic view of the fuel injection components and how they operate.*

1. *Fuel tanks*
2. *Tank changeover valves*
3. *Fuel pump*
4. *Fuel cooler (aircon cars only)*
5. *Fuel filter*
6. *Fuel rail*
7. *Fuel pressure regulator*
8. *Fuel injectors*
9. *Cold start injector*
10. *Air bleed valve*
11. *Non-return valve*

FLOODING PROTECTION
To prevent the engine being 'flooded' with fuel, a special circuit prevents the fuel pump from operating when the ignition is on except when the engine is being cranked or it is actually running.

OXYGEN SENSOR
North American models were fitted with a catalytic exhaust system which required an oxygen (or Lambda) sensor to be fitted in the manifold downpipe. Its function is to measure the oxygen/fuel ratio many times per second and pass the information to the ECU. When the mixture is too weak, the ECU puts more fuel to the injectors; when it is too rich, it reduces the amount of fuel, thus increasing the oxygen level. The ultimate aim is to achieve a fuel/air mixture of 14.75:1, a figure known as the Stoichiometric level. This is also denoted in scientific terms as the Greek letter Lambda – hence its adopted name.

BASIC CHECKS
If you have a multimeter, you can make several checks on your fuel injection system. WARNING – never connect a multimeter set to a resistance setting (Ohms) into a circuit which has power in it, as it could seriously damage the meter. Before making resistance checks it is wise to turn off the ignition. If you are making a check where the engine needs to be cranking, prevent it from starting by removing the negative (–) cable from the ignition coil.

COOLANT TEMPERATURE SENSOR (RESISTANCE – OHMS)
If the coolant temperature is reading incorrectly, this can affect the fuelling to the engine. Pull off the multi-plug and remove the switch itself. Connect the meter leads to the two terminals in the sensor. The resistance reading is related to the coolant temperature, getting lower as the temperature gets higher, as follows:

Resistance (Ohms)	Temperature (°C)
9.2K	-10
5.9K	0
2.5K	20
1.18K	40
600	60

The switch cannot be repaired and if faulty must be replaced.

THERMO-TIME SWITCH (RESISTANCE – OHMS)
Poor starting and idling is often a faulty thermo-time switch which is failing to activate the cold start valve – hence it is similar to starting a cold carburettor'd engine without the choke. Pull off the multi-plug and then remove the switch itself. This should be immersed in coolant (many owners repeat the old method of thermostat checking – that of heating a pan of water on the kitchen stove!) whilst the meter is connected between the body of the switch and terminal G. The time readings should match those given here.

Delay (secs)	Temperature (°C)
8	-10
4.5	0
3.5	10
0	36

When the meter is connected between the body of the switch and terminal W, the resistance figure should be extremely low. The switch cannot be repaired and, if faulty, must be replaced.

COLD START VALVE (VOLTAGE – APPROX. 12V)
Set your meter to VOLTS and connect to the wiring in the back of the cold start valve plug. When the cold engine is cranked, there should be battery voltage recorded. To further check, remove the injector and place in a suitable container so that no fuel can escape, and crank the engine. Fuel should spray into the container. The battery voltage and spraying fuel should both only last a few seconds. With a warm engine, there should be no voltage recorded. The valve cannot be repaired and, if faulty, must be replaced.

INJECTORS (RESISTANCE – OHMS/VOLTAGE – 12V)
Before starting work, make sure the whole area is absolutely clean and that no dirt can get into the manifold or the injector itself. Do not remove the injectors unless the engine is stone cold.

Remove each injector multi-plug and connect the meter across the two terminals on the injector. At an ambient temperature of around 20°C (68°F) you should get a reading of around 2.4 Ohms. With the multi-plug still removed, switch the meter to volts and, with ignition on, put the meter across the two contacts in the plug. There should be a battery voltage reading (around 12v). If not, it points to a break in the circuit somewhere, or possibly a blown fuse.

With a COLD engine and a depressurised system (*see earlier*) remove the injectors. Though the spray pattern can be tested on a DIY basis, it is better and safer to have the injectors checked for performance by a professional, preferably one of the Lucas/Bosch service centres. It is important that the injector pattern is smooth and even to get the best performance. According to David Abbott of injection specialists, Performance Car Services, unless the injector is very old and/or damaged, he will usually be able to clean it electronically to bring its performance back to standard. Better still, the checking, cleaning and re-checking process usually works out to be less than half of the cost of new injectors.

AIR TEMPERATURE SENSOR (RESISTANCE – OHMS)
Remove the battery earth cable and the multi-plug at the air flow sensor. By definition, the resistance across terminals 6 and 27 of the sensor will vary depending on the temperature. Compare with the following table:

Resistance (Ohms)	Temperature (°C)
9.2K	-10
5.9K	0
2.5K	20
1.18K	40
600	60

OVERRUN VALVE

For this check the engine must be running and so great care must be taken to avoid hot and moving parts.

Remove the valve (it's under the air distribution block) and the rubber hose between it and the throttle body. Start the engine and check the idle speed. If it is correct, then the idle valve is OK. If it decreases, the idle valve is faulty – it cannot be repaired and so should be replaced.

AUXILIARY AIR VALVE

The auxiliary air valve should only be operational when the engine is cold. Start the engine (cold) and, taking the usual care, gently squeeze the rubber hose linking the valve to the inlet manifold. The idle speed should drop – if it does, it shows the valve to be working properly. With a hot engine, repeat the test, at which point it should have no effect at all. If the valve fails either of these tests, it should be replaced.

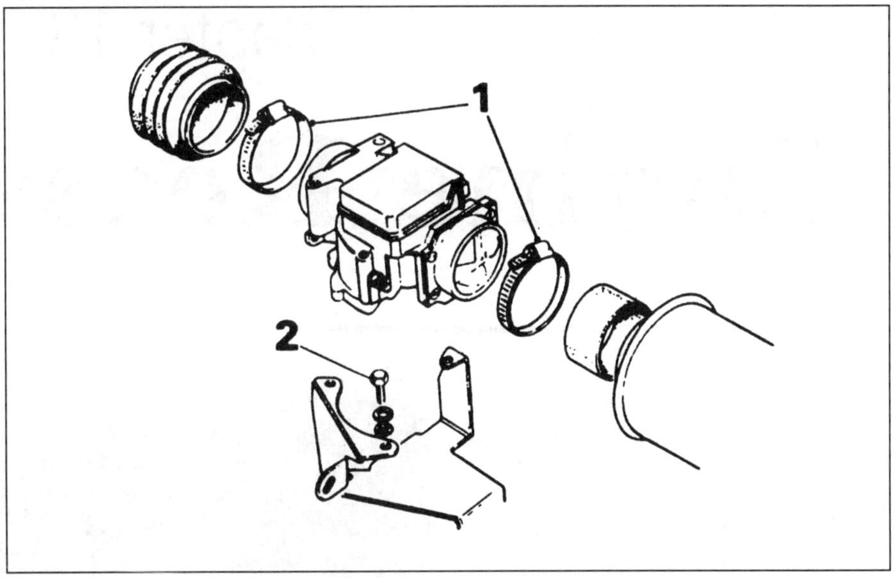

▲ *This diagram shows the component parts of the air flow meter. Undoing one bolt and loosening the Jubilee clips on the flexible air connections is all that is required.*

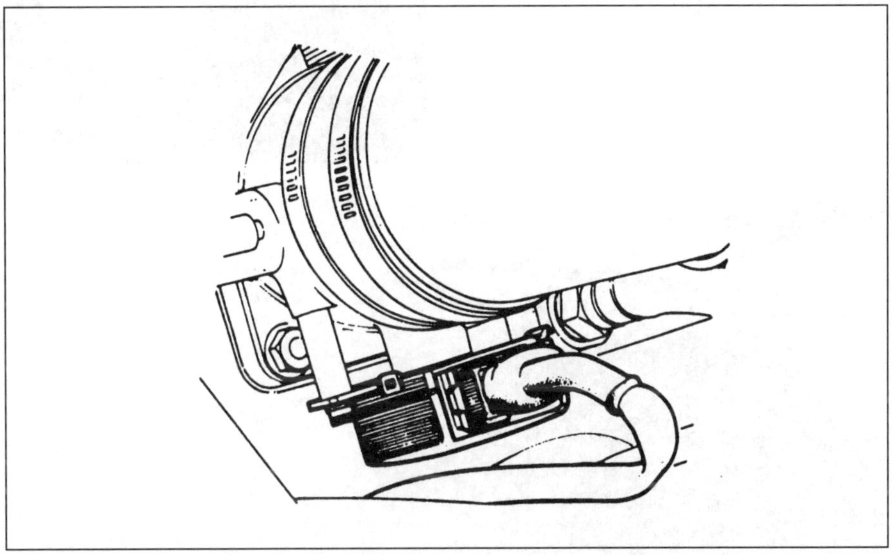

▲ *Two types of throttle switch were fitted. This is the type which is operated directly by a spindle and ...*

◀ *... this type is operated by a lever, seen at the top, which moves the flap as in a carburettor.*

Chapter 18

Engine Reassembly

If you are rebuilding your own engine it will probably take weeks or even months to reach the stage of reassembly. It's at this stage you'll be glad you carefully kept both records of which bits go where during the stripdown process and all the various re-usable fasteners in suitably marked boxes (tins have to be used with some care, as they can cause steel fasteners to rust).

▲ The rear circular oil seal cover is a bone of contention with most XK owners, it being a common source of oil leaks and a darned nuisance to replace. The rope seal is in two halves which locate by a hole/dowel at each side. Tim fitted the seal into one half, 'rolling' it firmly into the groove by using a steel shaft, and then cutting off any excess with a sharp blade. To prevent the seal moving around in the cover, a small amount of PU adhesive is used. The seal was sized using a special VSE tool (but the same as Jaguar used on the production line) in an old XK block.

◀ The seals and covers were soaked overnight in engine oil.

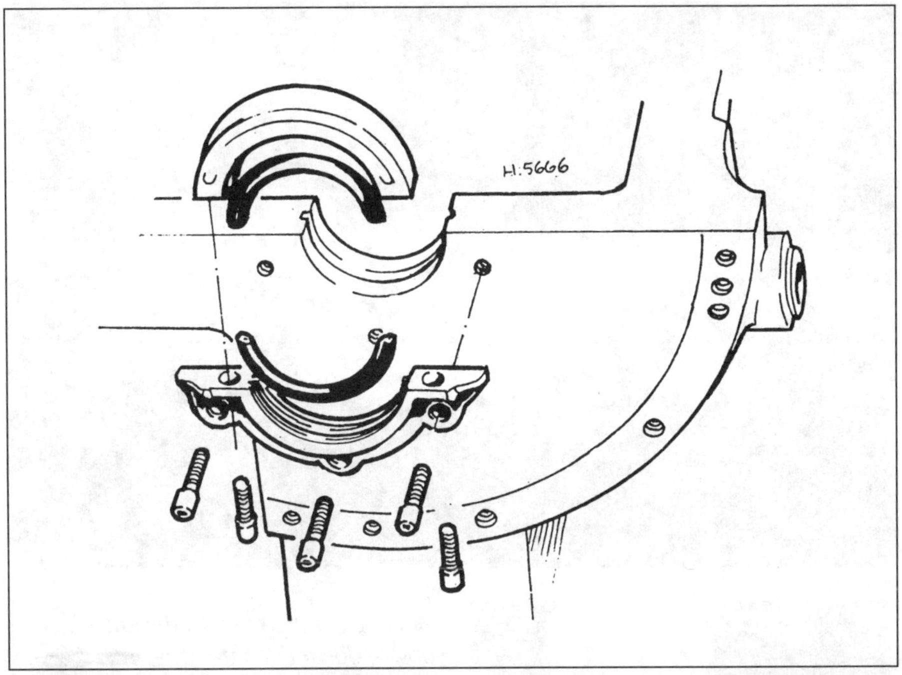

◀ *This shows how the crankshaft oil seal assembly fits.*

◀ *Rob Beere Racing Services has developed a standard lip seal conversion which can be applied to all XK engines. Because it's a standard, single-piece seal, it's cheaper and far more efficient than the rope version. Moreover, it can be changed without the need for taking the engine out. As can be seen here, the end of the crank has to be machined down to suit the new seal. A standard crank is seen here being held against a lip seal conversion in situ.*

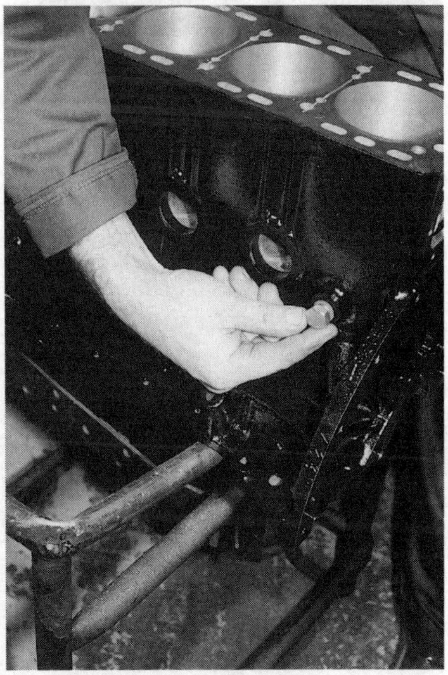

▶ *Meanwhile, Malcolm prepared the block. He replaced the front oil gallery plug/washer, the four side plugs/copper washers, the large rear hexagon plug and the water drain plug (some models have a tap here).*

▶ *He put the tiny cone filter in place – this will ultimately be behind the hydraulic tensioner.*

◀ *Malcolm then turned the engine upside down and fitted the lower part of the crankshaft oil seal, using Hylomar universal blue gasket compound.*

▶ *Malcolm squirted a little oil into the galleries before putting the main bearing shells into place. Graphogen colloidal graphite grease was smeared on to the inner surfaces to provide some start-up protection for the bearings.*

◀ *He carefully placed the freshly balanced and cleaned crankshaft in position. It's important that the crank should not be knocked or chipped whilst it is being manhandled into place, and it should be laid fair and square on to the bearing shells.*

▶ *For easy extra power gain (and a slightly smoother engine) RBRS have this crankshaft scraper plate. As the name suggests, it fits into place in the sump and scrapes off the oil which tends to stick to the crank as it revolves, thereby allowing the crank to turn more easily and quickly – effectively making it lighter. You could expect to gain anything between 5 to 8bhp on a rebuilt engine.* (Courtesy Rob Beere Racing Services)

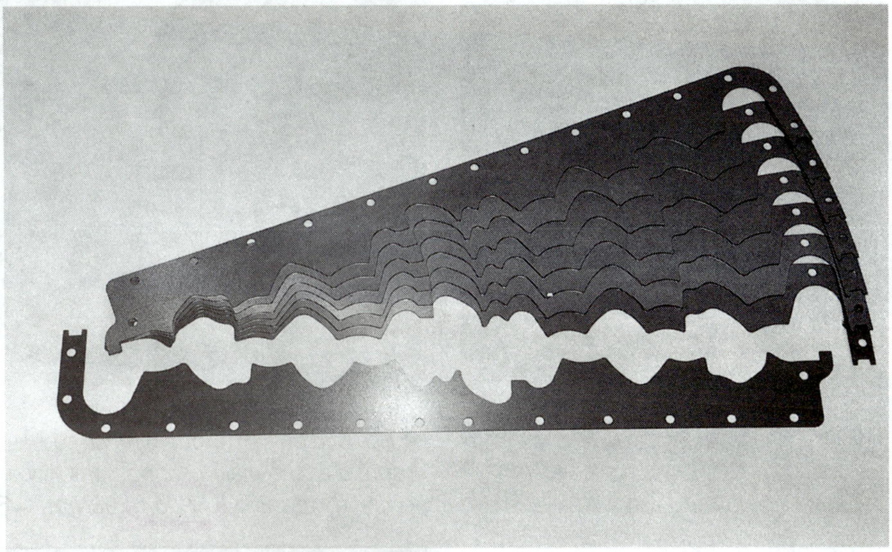

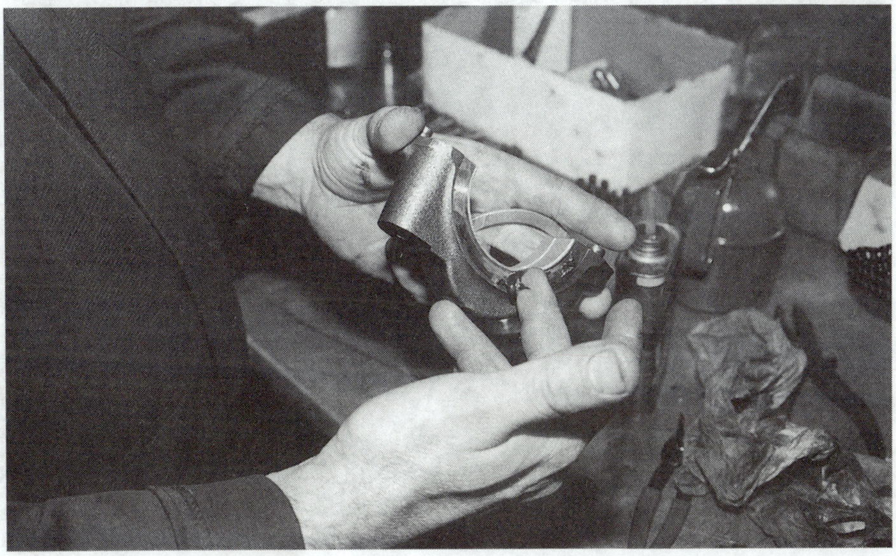

◀ *The crank shells were put into the main bearing caps 'dry', but Graphogen grease was again used on the inner surfaces and on the sides of the centre bearing cap thrust washers. Thrust washers should be fitted according to the amount of endfloat in the crank (see later) and, if required, it is permissible to use standard and oversize at either side of the centre bearing cap.*

▶ *The main bearing caps have to be fitted with the numbers aligned correctly – the block is stamped 1–6 and so are the caps.*

◀ *He used an S-P torque wrench to ensure that the main bearing bolts were accurately torqued to 72lb/ft (98Nm).* - 83 on others

▶ *The top half of the crankshaft oil seal was fitted, again using Hylomar universal blue gasket compound and a test dial indicator to check the crankshaft endfloat. This should be between 0.004 and 0.006in (0.10mm–0.15mm). He also checked that the crank rotated freely and that there were no 'tight' spots. Each main bearing bolt has a tab washer, which has to be bent back, as shown here. Those for the rear main bearing bolts being longer than the others – the plain ends should be tapped down round the bolt hole bosses.*

◀ *Each con rod bearing shell was fitted 'dry' and Graphogen grease applied to the inside of the shell. The bores were all liberally oiled, as were the piston rings.*

▶ *There's an arrow stamped into the top of each piston and this MUST face towards the front of the engine. In this picture of a 4.2-litre E-type engine the arrow is plain to see, as is the reference number (at 6 o'clock), and at the 12 o'clock position is a marking denoting that this particular piston is '+.020', ie 20 thou oversize, following a rebore.*

◀ *The con rods are numbered and must be fitted in the correct order. An S-P ring compressor was used to get the pistons into the bores. It's virtually impossible to do this without a compressor, with the very real risk of snapping the rings and/or damaging the bore. Once the piston is squarely in the bore, use a wooden hammer handle to tap it gently down, making sure the crank is turned to a suitable position.*

▲ The con rod caps were fitted and Malcolm applied a dab of Loctite 243 thread locker as a belt and braces measure to ensure the bolts don't work loose. Once more, it is important that the bolts are torqued up correctly. Note that the figure of 37lb/ft (50Nm) is around half that required for the main bearing cap bolts, and really isn't much twisting power at all. Malcolm put Nos. 1 and 6 pistons at TDC (top dead centre) and went to the front of the engine and …

TIMING CHAIN ASSEMBLY

▲ … fitted the lower timing chain guide followed by the intermediate guide – it's important not to tighten the securing bolts at this stage.

▶ The complete timing chain assembly was fitted, and the bolts were tightened.

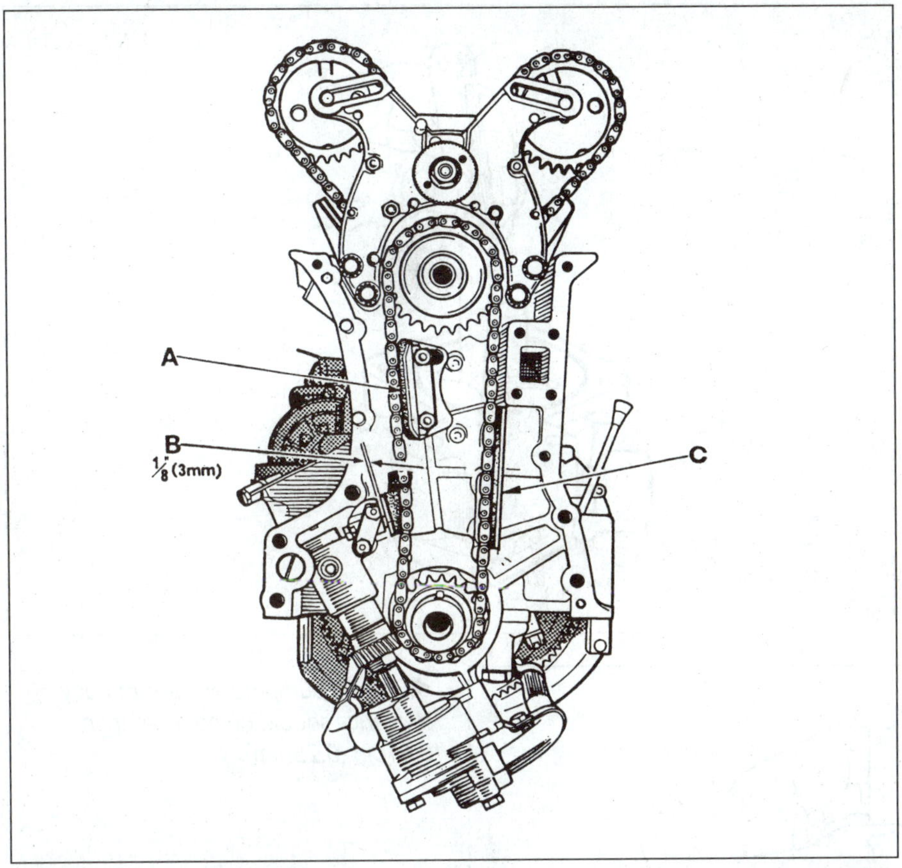

◀ *It's important to grasp what goes where when considering the timing chain layout. This diagram shows a 4.2-litre E-type. The intermediate tensioner was tightened and the tabs on the locking plate prised over. The hydraulic tensioner was fitted and adjusted to have around ⅛in (3mm) play, as detailed in this diagram – remember that the engine shown in the sequence here is upside down on the workbench. Again, all the locking tabs were prised over.*

A Intermediate guide
B Slipper-to-body clearance
C Lower guide

▼ *The two timing cover gaskets were put into place, using a smear of PU adhesive to keep them in place until Malcolm fitted the timing cover itself. He also fitted the front oil seal, using Hylomar compound and smeared grease around the front of the crank.*

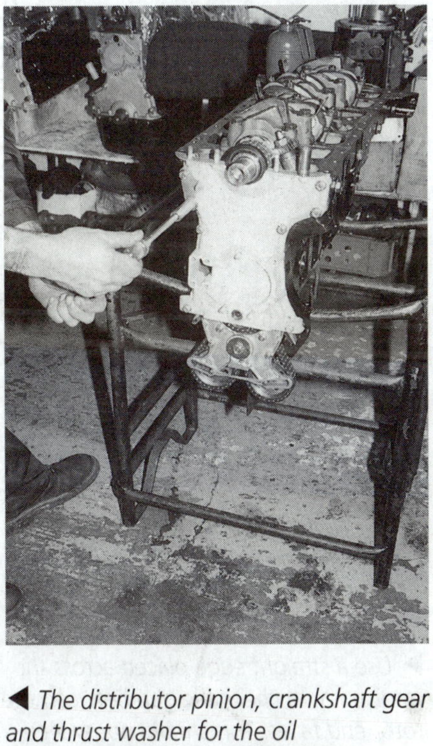

◀ *The distributor pinion, crankshaft gear and thrust washer for the oil pump/distributor driveshaft were installed. The gear locates in a Woodruff key on the driveshaft. Here, Malcolm is tapping over the locking tab.*

OIL PUMPS

▶ *Having removed the oil pump, it should be opened up by removing the four bolts from the pump body. Take out the rotors but do not remove the inner rotor from the driveshaft. If either the rotors or pump body show signs of scoring or heavy wear, the most practical answer is to replace the oil pump complete. If everything looks OK, you will still need to check to ensure that the rotor clearances are within the tolerances specified. Use a feeler gauge to check that the clearance between the inner and outer rotors is no more than 0.006in (0.15mm).*

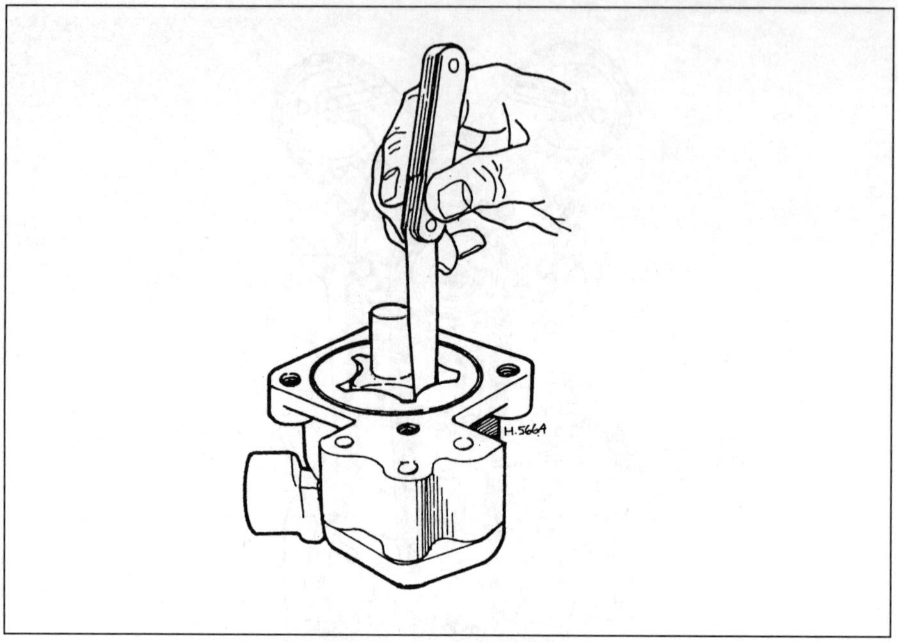

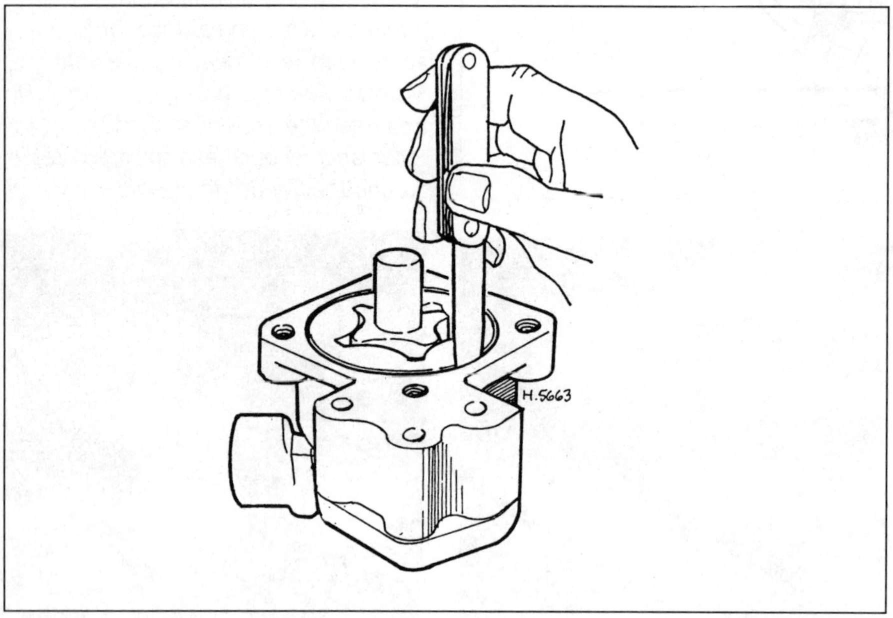

◀ *The oil pump outer rotor to body clearance should be no more than 0.010in (0.25mm).*

▶ *Use a straight edge placed across the rotor body to check the gap between the rotor end faces. This should not exceed 0.0025in (0.06mm). See the stripdown section for an exploded view of the oil pump and distributor driveshaft assembly.*

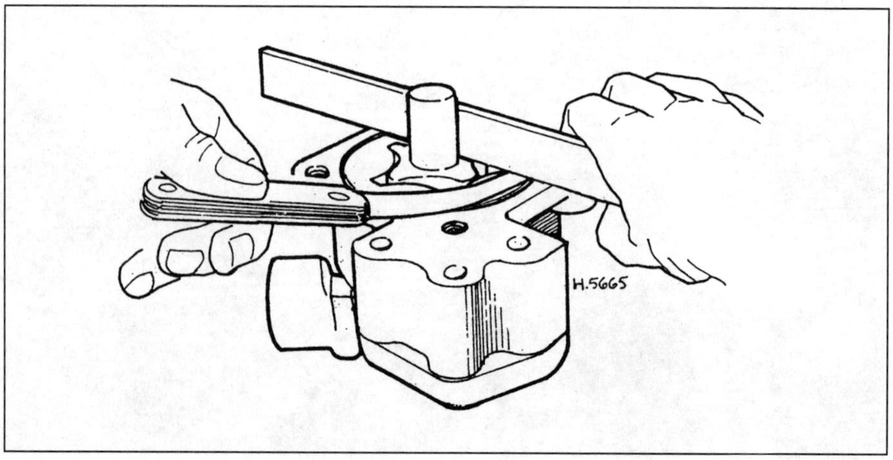

◀ *This photo shows the cleaned-up and checked pump body with the spring and plunger mechanism. A VSE tip is to always ensure that there is a washer at the end of the spring opposite the plunger; many DIY builders forget this small item – with big (expensive) results! The end cap gasket was smeared with Hylomar compound. Malcolm fitted the bush on to the distributor driveshaft, and added the oil pump. Assuming the clearances are OK, VSE always fit new O-rings. It was primed by filling it with oil. The three securing bolts all have locking tabs which have to be bent back.*

▶ *With effect from the XK 140SE models, all oil pumps were rotor type, having previously been the gear type. This photo shows the two main types of pump fitted since then. On the left the small rotor variety fitted up to 1962/63, and on the right the large rotor type fitted from then onwards.* (Courtesy VSE Engineering)

◀ *Compare with the gear types fitted prior to the XK 140SE. On the left is an XK 120 and on the right is a pump from an early (non-SE) XK 140.* (Courtesy VSE Engineering)

▶ *It was not only the oil pumps that changed over the years. This photo shows the various types of oil filter heads fitted to the XK engine.* (Courtesy VSE Engineering)

◀ *For any uprated engine, attention to the oil pump is a must. Rob Beere Racing Services has developed a pump (seen here on the right, alongside a standard unit) which, by virtue of its raised rotor, has a 40 per cent increase in oil throughput.*

▶ *Naturally, the oil pump top has had to be machined to take the extra height of the new rotor. Note the increased depth of the top on the right (the uprated version) compared to the standard one on the left.*

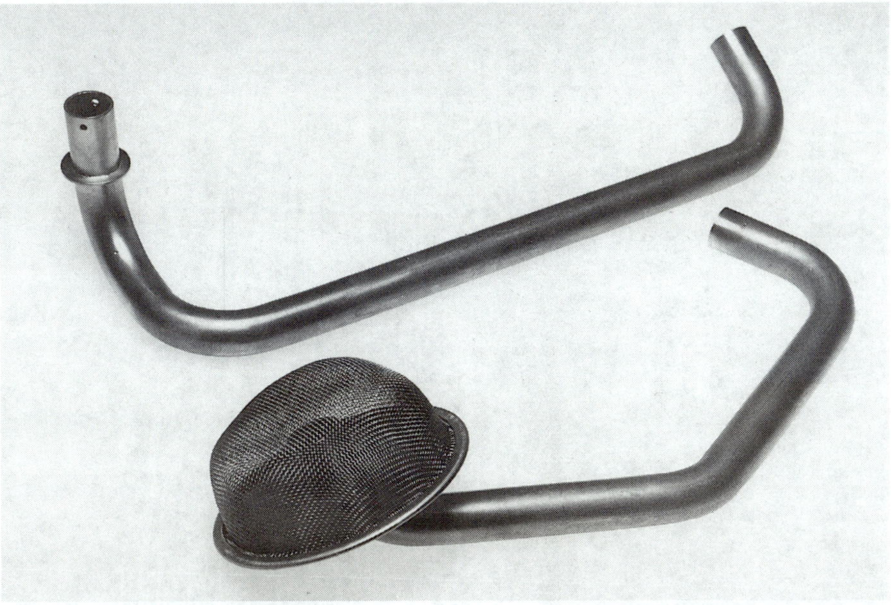

◀ *If you have a bigger bore oil pump, or are using a later type pump with an early engine, it's a good idea to use larger bore oil pipes, as seen here.*

▲ *According to Carl Taylor from RBRS: 'It's always worth fitting an oil cooler to an XK engine when it's tweaked in any way. We offer two types of filter kit – one vertical with a screw-on canister, and the other is horizontal. The pipework differs from road engine to race engine; not surprisingly, the race engine needs bigger pipework. Many tuners advise the use of thermostats, but we don't, because from our experience you'll simply never over-cool an XK engine.'*

The size of the oil cooler depends very much on the power output of the engine and its intended use – again, race cars require larger coolers than pure road cars. If you fit an oil cooler in front of the radiator (as on this RBRS racing E-type) don't leave a gap behind it, even a small one. What tends to happen is that the momentum of the air rushing through the oil cooler is lost and it doesn't blow through the water radiator as it should. The result is a hot spot on the radiator the size of the oil cooler, and decreased engine cooling ability. RBRS always fit a rubber seal between the cooler and radiator to ensure that the cooling air passes straight through both sets of fins with its speed more or less undiminished.

▲ *Malcolm then fitted the two oil pipes into position, greasing the ends of the pipes where they went into the oil pump to make them an easier fit. Don't forget the triangular gasket where the oil pipe goes into the crankcase – again Hylomar compound was used to make a really good seal. Note that all three pipe brackets are different. The sump gasket was laid carefully in place. It's a large gasket and, because there are 26 holes, it is distinctly flimsy. To aid location, Malcolm put the studs back into their holes.*

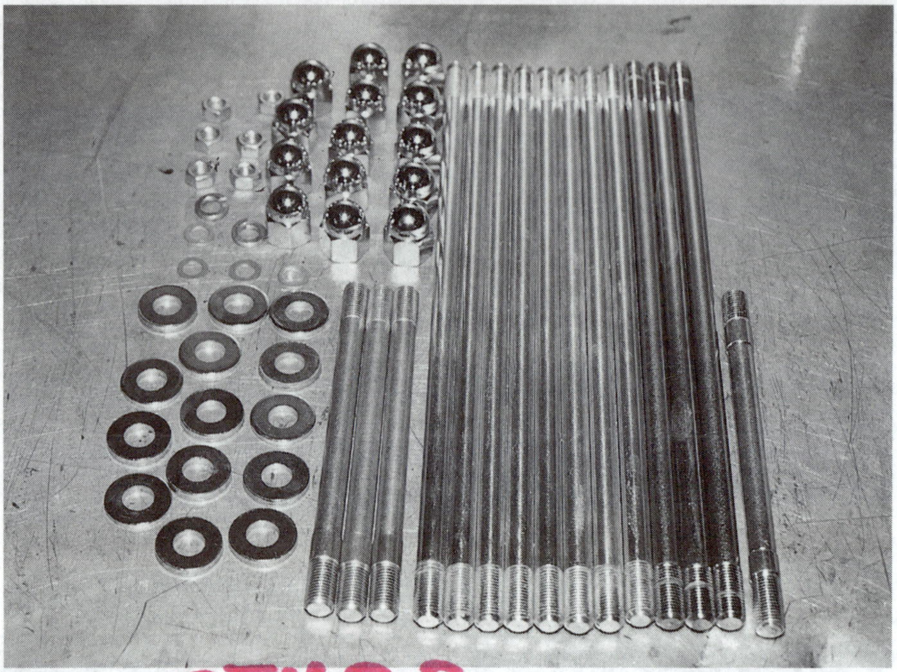

big baffle in sump STUDS

▲ He fitted a new sump plug and copper washer, the baffle and the oil seal before putting the sump carefully into place over the gasket. It's important to tighten the sump bolts in a cross-hatch sequence a little at a time to avoid distorting the sump (and they don't need to be wheel-nut tight, either!).

▲ After replacing the cone and damper on the crank, Malcolm started work on fitting the cylinder head, prepared elsewhere by Doug Whittingham. VSE always use new stubs, bolts, washers and nuts. As you can see, the studs are of different lengths, and it's important to get the right stud in the right hole. The studs here are plated and made specifically for VSE, the object being to make them much more corrosion resistant. As we've already seen, badly corroded studs are common, and if they seize in place it usually means a wrecked block. Because the company buys in such large numbers, the cost to their customers of these uprated studs is actually cheaper than standard studs from Jaguar.

▶ Malcolm applied copper grease to the threads going into the block before threading them in by hand and then tightening by using the S-P stud remover. Remember that he'd already chased through all the screw threads on the block to ensure that there would be no problems in this area.

◀ *Seen on the bench is a standard 3.8-litre head gasket. Note that it is of the earlier persuasion with only two waterway holes at the rear of the gasket. Compare with the gasket above it, which has four waterway holes. RBRS had a few problems with owners blowing head gaskets, especially when the engine had been tuned, so they had a special variety made by Cooper Payen (the company which makes the originals) although the cost is considerably more than the original. However, when mixed in with the general costs of tuning and uprating, it isn't that much – and it's even less if, otherwise, the head has to come off soon after the rebuild to replace the original gasket. Although the new gasket features four waterway holes, it can be used with earlier blocks and heads simply by cutting off the piece not required.*

◀ *The gasket used here is composite, so doesn't require any sealant, but steel-shimmed gaskets do. Make sure the word 'top' on the gasket faces upwards. Malcolm turned the crank so that No. 6 piston (the one nearest the crankshaft pulley) was at TDC. This can be confirmed by checking that the 'O' mark on the crankshaft damper is opposite the pointer on the crankcase. When fitting the new head, patience is required to ensure that neither head, block nor studs are damaged. It is vital at this stage that the mating surfaces are clinically clean.*

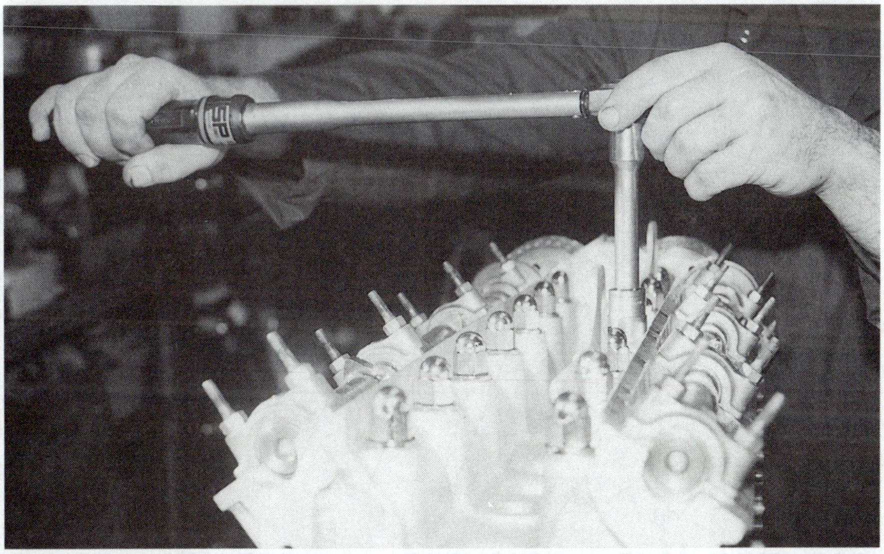

◀ *Each stud was oiled before the washers and new chrome nuts were fitted. Don't forget to refit the engine lifting lugs. Note, once again, the use of the S-P torque wrench – it's so easy to get carried away and ruin all that work.*

soft washers

▶ *When dealing with the six nuts securing the front of the head, note that the harder spring washers go on the back two studs where they tighten up to the cast-iron block. The front four tighten up to the alloy head and so have softer copper washers fitted. The cardboard insert in the distributor drive hole prevents dirt getting into the engine.*

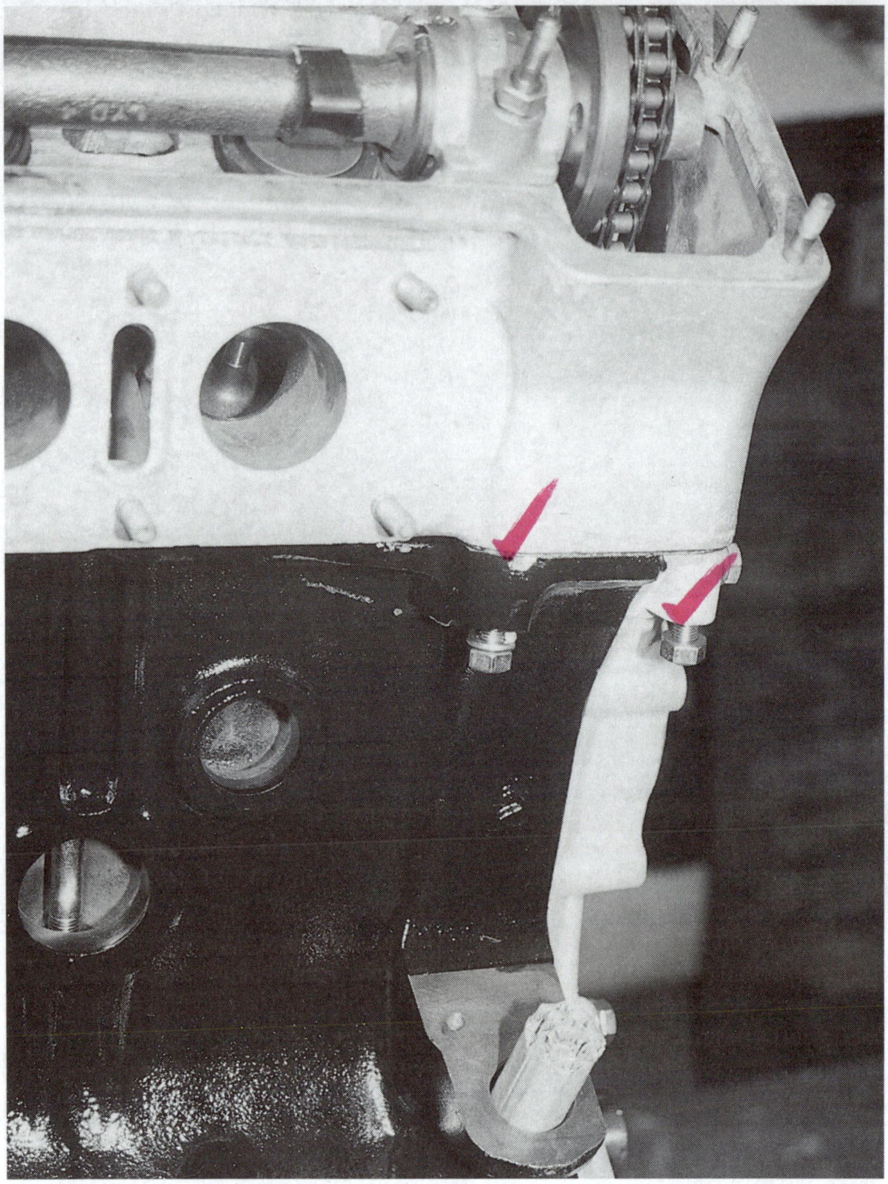

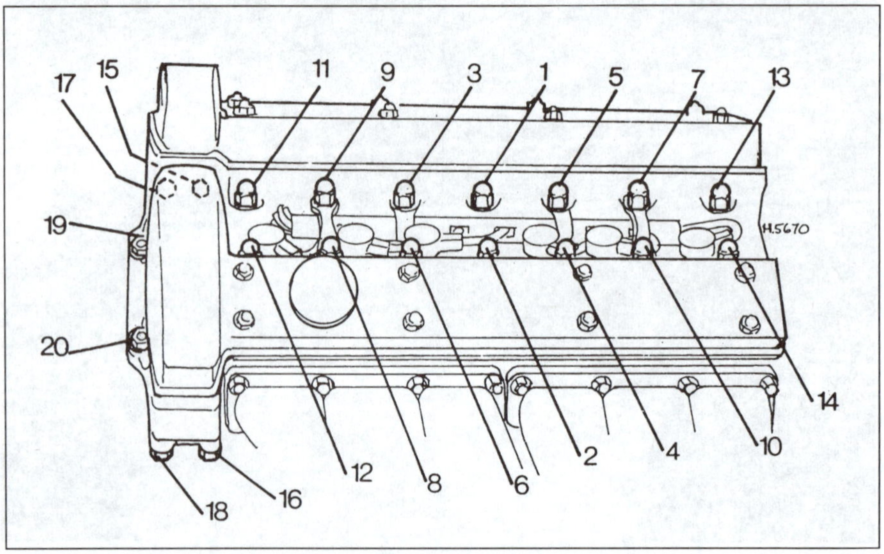

▶ *The cylinder head bolts should be tightened a little at a time in the sequence shown here. It is important to take your time here, and remember that too much torque could easily warp the alloy head; make sure that the head is tightened down square.*

VALVE TIMING

▲ *With No. 6 piston at TDC (as described earlier, make sure the notches in the camshaft sprocket are at 90 degrees to a line touching the two rocker cover mating surfaces of the cylinder head). The adjusting plates should be pulled forward until the serrations disengage (removing circlips if already fitted). Fit the sprockets to the camshaft flanges and align the bolt holes – if they don't align, then turn the plate through 180 degrees. Fit the circlips to the sprockets, fit the two bolts and turn the sprocket until the remaining two holes are showing, and fit the bolts in them. Remember that each pair of bolts must have the locking tab plate fitted which has to be knocked over the bolts once they're tight.*

▲ *Tension the chain by inserting the stop peg (plunger) inward and rotating the serrated plate anti-clockwise. You'll need the special tool, as shown here. You can buy this or hire it from one of the clubs. Don't forget to fit the circlips on to the sprockets.*

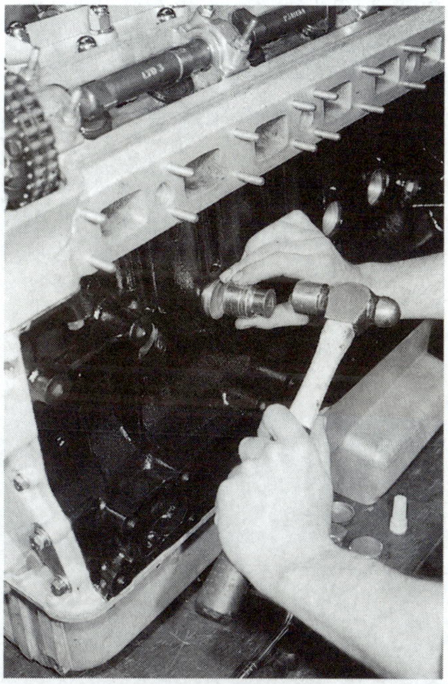

▲ *The cylinder head studs are visible through the core plug holes. Malcolm squirted Loctite 648 on to the core plug holes in the block, before tapping the core plugs home using an impact socket just a little smaller than the plug itself to ensure that it went in exactly square.*

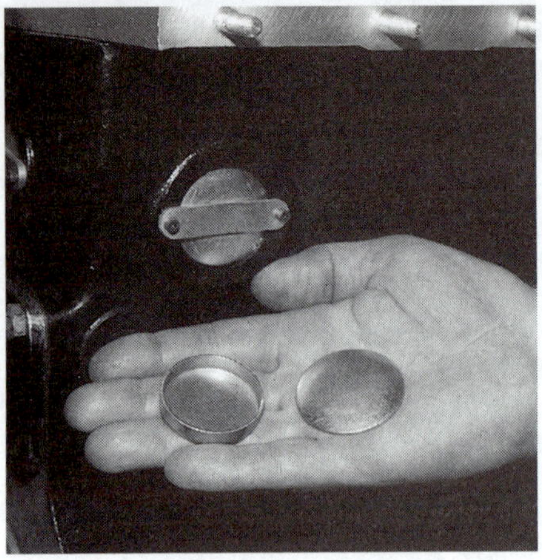

▲ *The core plugs used on the XJ6 are the later type cup plug. Compare this type (seen on the left here) with the earlier saucer type. The latter has an irritating (and expensive) tendency to blow out under pressure. To prevent this happening RBRS fits a strap across the plugs which is secured at either end by a small Allen headed setscrew. The company did it so often that it now includes it as part of the service on offer or as a small kit for DIY fitting (when it includes the small tap required for the securing screws).*

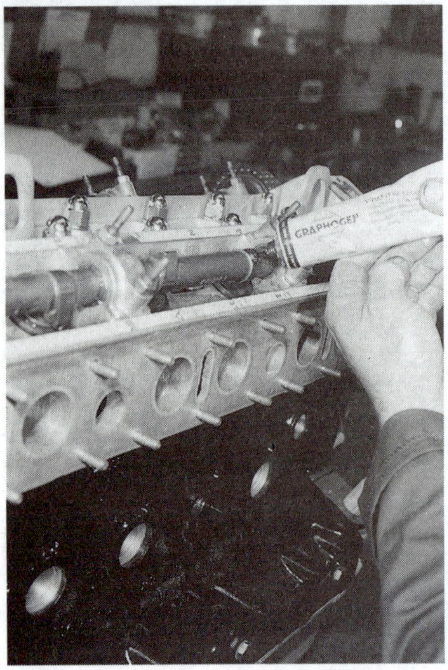

▲ *The rear water plate cover was refitted, and then the cam lobes were all given a smear of Graphogen grease to prevent them starting 'dry'.*

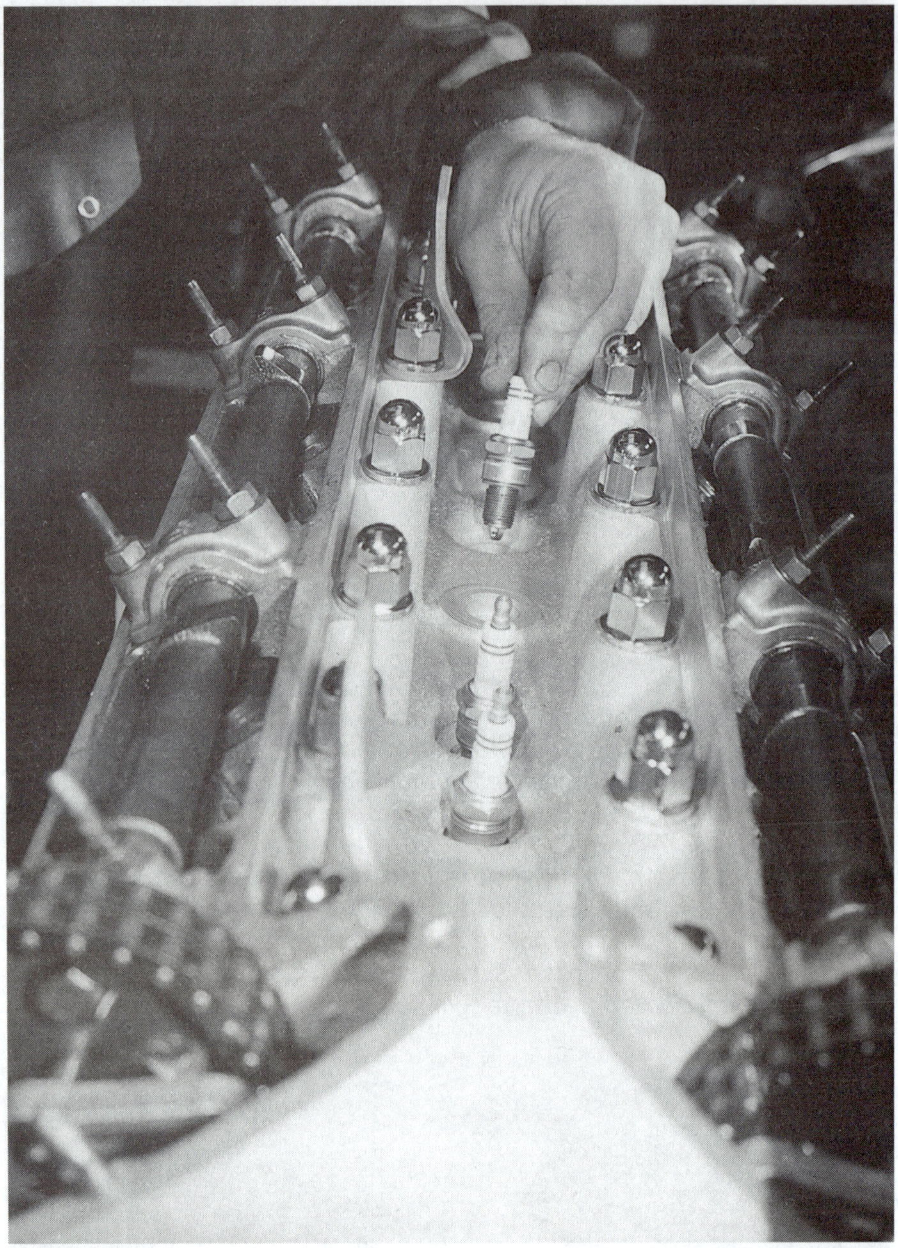

▲ *A new set of plugs was fitted. Spark plugs are a greatly neglected part of most engines, and yet they have to work in atrocious conditions, dealing with tens of thousands of volts in temperatures upwards of 700° for hours on end. On the one hand, they have to be 'hot' enough to allow easy starting and low-speed operation, and on the other they have to stay 'cool' during sustained periods of high engine speeds. If ever there's an unsung hero, it's the spark plug! All spark plugs should be tightened to slightly more than hand tight, around 18lb/ft (25Nm). Where you have changed an engine specification, it's important to ensure that the original spark plug specification is still suitable. Apart from not performing as it should, a wrong grade of spark plug could actually cause engine damage. Refer to your engine specialists or directly to the spark plug manufacturer.*

Early XK 120 engines can suffer from electrical problems because of the routeing of the plug leads. Initially, they were routed the long way round, under the carburettors to the rear of the head, where they can short-out. Because of this the factory started re-routeing them to the front of the head.

◀ *Malcolm fitted the rocker cover gasket, using Hylomar on the metal surfaces of the head and the rocker cover. It's important not to over-torque the rocker cover nuts. Remember, it's a fairly light aluminium casting and very easily damaged. The oil-sealing capability must come from the gasket and Hylomar.*

▲ *A nice final touch is a new Jaguar badge.*

▲ *Every engine VSE produces is thoroughly road-tested on this purpose-built rig. All aspects of engine running are checked, including timing, oil and water leaks and water temperature. VSE export engines all over the world, so they simply HAVE to be right when they get there. Note that, although this is a fuel-injection engine, it was the work of minutes to bolt two SU carburettors on to the side for testing purposes.*

▶ *In truth, every serious XK engine concern should have (or, at least, have access to) a Dyno facility. This applies no more so than with tuning, where extra power/torque is expected. Without one, the tuner can only be guessing. This a tuned E-type engine, complete with triple Webers, on the Rob Beere Racing Services Dyno rig.*

RUNNING IN YOUR NEW ENGINE

Having spent all that time and money building up your perfect engine, it would be a shame (to say the least!) to blow it up within the month. Tim Camp of VSE recommends the following regime as the prelude for a long and happy life – anything up to 150,000 miles – with your rejuvenated XK:

SCHEDULE

Make sure that the correct amount of quality oil is in the engine and that the cooling system is sound, operating correctly and has the correct level of anti-freeze (it should contain corrosion inhibitors).

For the first 500 miles, do not exceed 3,000rpm. Do not labour the engine and avoid constant speed driving situations (ie motorways). Vary engine speeds as much as possible up to the rev limit. It's generally advisable to use a fairly average 20/50 oil (ie not a synthetic) as this will tend to allow the various contacting moving parts to bed-in properly. The more esoteric oils are so good at lubricating, that they don't allow this all-important process to go through.

500 MILES

Drain the engine oil, fit a new filter and refill with a quality oil. With engine COLD, re-torque the cylinder head nuts to 60ft/lb (82Nm). Check all hose clips, security bolts, etc, look around the engine and the engine bay for water/oil leaks. Check the ignition timing and the fuel mixture – this has to be carried out using an exhaust gas analyser, either at your local garage or using the Gunson's Gastester.

500–1,000 MILES

Gradually increase the maximum rev limit from 3,000rpm to 4,500rpm for short bursts.

1,000–1,500 MILES

Build up to maximum revs for your particular model. Do not use sustained high revs for long periods.

1,500 MILES

Repeat service instructions as per 500 miles.

The running-in period is now complete and the service intervals should be as the factory recommendations for your model.

Tim is keen to stress that Jaguar engines suffer most wear and tear on short journeys – those less than 10 miles in total being particularly harmful. Make regular checks on the cooling system, anti-freeze content, ignition timing and carburation. If you do make mostly shorter journeys, make more frequent oil changes than recommended by the factory.

Chapter 19

Tuning

(All the photographs in this section were provided by Rob Beere Racing Services.)

How deep the sea, how high the sky? Tuning the XK engine is an occupation limited largely by the size of your wallet and the extent of your imagination. However, for practical road-going purposes, it has to be remembered that the XK is – or should be – no mean performer straight out of the box. And that's where the problems start, because the XK-powered car you buy will probably be a long way from its original box and have a rather old and often ill-serviced engine 'neath its bonnet. So, as ever, the first step is to level the playing field – there's little point in whacking huge Weber carburettors atop an engine with a leaking cylinder head and a cracked crank!

◀ *Not your average installation! This 3.8-litre XK is clearly installed in an E-type, but is equally clearly very non-standard. Note the purpose-made alloy air-cleaner housing feeding those three thirsty 48 DCOE Weber carbs, and the alloy radiator. The car belongs to an extremely enthusiastic Swedish RBRS customer, who has spared no expense to make his E-type slightly quicker (!) than normal. The engine is currently (he uprates it on a regular basis) producing 325bhp and a gut-wrenching 350ft/lb torque.*

▶ *Polished, pristine and powerful – another Beere-tuned 3.8-litre unit, once more with triple Webers providing the fuel and air mixture. They're attached to a replica wide-angle head, and the unit is destined for a replica Lightweight E-type (the prices and scarcity of the originals make it no sin to replicate the car). To be totally original, mechanical fuel injection should have been used, but the cost was eye-wateringly prohibitive – even in racing terms! Note the uprated horizontal canister oil filter. As you can see, the sump looks unusual and …*

The very first task is to strip down the engine (as shown previously) to find out exactly what you've got and, more important, what you haven't. Simply 'blueprinting' the engine – ie building it *exactly* to the original manufacturer's specifications – will result in a very pleasant unit indeed. Getting the crank exactly in balance and the pistons exactly the same weight are the sort of details that no production line can expect to achieve. And, of course, replacing old and worn components never goes amiss.

Obviously, if tuning is your aim, it's best to consider the whole package whilst the engine is in kit form on the bench; there's little point in rebuilding it only to decide later that a crank upgrade is required. And it's no use going in search of extra power and torque if you're not going to upgrade the bottom end. If you're short of ideas, check out the mouth-watering menu of tastily tuned XK engines in this section.

▶ *... even more so when the engine is upended. This is a dry-sump, a fact made clearer ...*

▶ *... with the sump removed altogether, revealing the twin oil pumps used to distribute the oil to the engine via a remote reservoir.*

▶ *This 3.8-litre E-type full-race engine has been built for Historic GT racing. Whilst Jaguar coloured the cylinder heads (or deliberately left them polished silver) little attention was paid to the blocks. However, the owner of this car has had his block sprayed blue – to match the car! Carburation is via three SU HS8s (Webers not being allowed in this class) gulping in air through those lovely polished bellmouths.*

◀ *Looking into the (usually) oily bits, we find an uprated crank with knife-edge webs (to cut through the oil better) and those exquisitely engineered – and priced – Carrillo con rods.*

▶ *With the bottom end almost finished, we can see the larger oil pipes and the, not so obvious, internally uprated oil pump.*

◀ *Another interesting toy – a full race engine stroked out to a healthy 4.6 litres, producing around 315bhp and 350lb torque. You can see that the lump tagged on to the back is not your normal 3-speed auto – its a Getrag 5-speed unit.*

▶ *Most XK uprating starts with attention to the cylinder head. Polishing, porting and increasing the valve sizes all helps increase the power and torque. This is a block's-eye view of a 4.2-litre, S2 E-type head complete with a tasty trio of SU carburettors.*

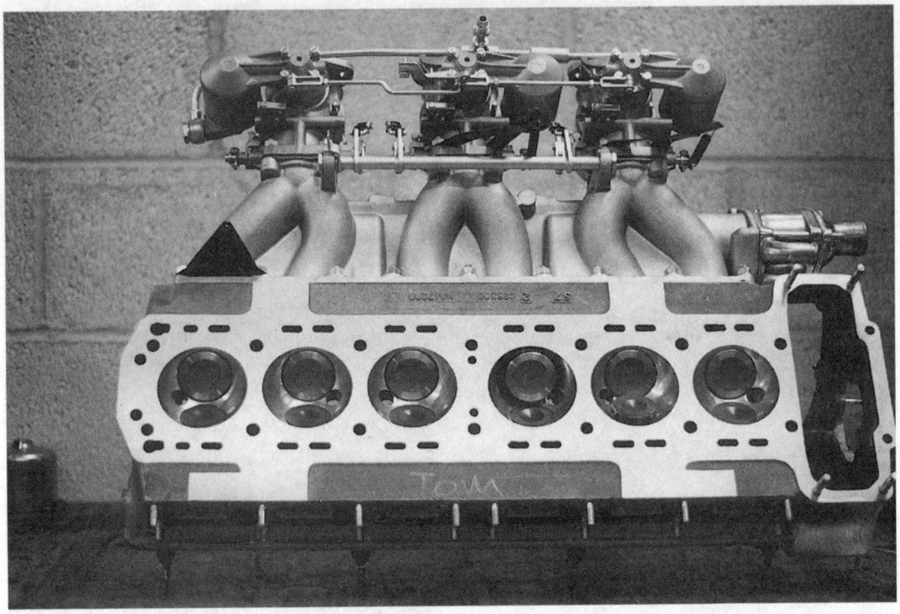

▶ *Serious power almost always means Weber carburettors. They are superbly engineered pieces of equipment, but much more complicated than the SUs that most XKs come with as standard. However, there's nothing for nothing, especially when it comes to engine tuning.*

Appendix

SPECIALISTS AND SUPPLIERS

AG Products
Manufacturers of the Battery Sava device.
PO Box 575, Ascot, Berks SL5 9XS
01344 24131

Autotune (Rishton) Ltd
Manufacturers of Aristocat sports cars using 60–90 per cent Jaguar parts, including XK engines.
Unit 1 Riverside Industrial Estate, Rishton, Blackburn, Lancs BB1 4NF
01254 886819

Barratt, SNG
Suppliers of all XK repair and maintenance parts, including all engine components and most Jaguar car parts as well.
The Heritage Building, Stourbridge Road, Bridgnorth, Shropshire WV15 6AD
01746 765432

Rob Beere Racing Services
Suppliers of XK engines for concours-winning Jaguars, racing series across the world, including the Bridgestone Challenge, German Historic GT and Post Historic Touring Cars – not to mention the world's fastest E-type! Instrumental in the compilation of this book.
Unit 9, Herald Business Park,
Golden Acres Lane,
Binley, Coventry CV3 2SY
01203 650521

Broomstick Cars Ltd
Manufacturers of the BS120 Roadster, an XK 120 look-alike using many Jaguar components.
Willow Farm, Ivinghoe Aston,
Leighton Buzzard, Beds LU7 9DG
01525 220123

Burlen Fuel Systems Limited
SU/Stromberg carburettor sales and spares.
Spitfire House, Castle Road,
Salisbury, Wilts SP1 3SA
01722 412500

Carburettor Exchange
Carburettor specialist.
28F High street, Leighton Buzzard,
Beds LU7 7EA
01525 371369

Clarke International
A wide range of restoration equipment including MIG/ARC welders, compressors and pneumatic tools, hoists, grinders, drills, etc.
Hemnal Street, Epping, Essex CM16 4LG
01992 565300

Draper Tools Ltd
Suppliers of all kinds of hand tools and distributors of Metabo power tools.
Hursley Road, Chandlersford,
Eastleigh, Hants SO5 5YF
01703 266355

Falcon stainless steel exhausts
See SC Parts

Jaguar Cars Limited
Manufacturers of Jaguar cars and of original equipment spares.
Browns Lane, Allesley,
Coventry, CV5 9DR
01203 402121

Jaguar-Daimler Heritage Trust
Repository of Jaguar's heritage, a treasure trove of XK history – and cars.
Browns Lane, Allesley,
Coventry CV5 9DR
01203 202141

K & N Filters (Europe) Ltd
Range of long-life air filters for better breathing and improved torque and power.
Wilderspool Causeway, Warrington,
Cheshire WA4 6QP
01925 636950

Kenlowe Limited
Electric fan replacement kits and Hot Start warm-up device, both designed to aid efficiency and improve mpg.
Burchetts Green, Maidenhead,
Berks SL6 6QU
01628 823303

Le Mans Sports Car (1996) Ltd
Makers of Jaguar engine-based kit cars.
Ford Farm, Uffeulme, Cullompton,
Devon EX15 3DQ
01884 841927

David Manners
Wide range of tuning and maintenance parts for all XK engines.
991 Wolverhampton Road, Oldbury,
West Midlands B69 4RJ
0121 544 4040

Mr Fast'ner
A vast range of fasteners and 'Recoil' thread repair system.
Units 6/12, Warwick House Industrial Park,
Banbury Road, Southam,
Warks CV33 0PS
01926 817207

Northampton Autorads
Established in 1968, specialists in repair, rebuilding and manufacture of all types of radiator, air-conditioning condensers and water wipes. Supply of all Jaguar radiators world-wide and O/E radiators to Aston Martin Lagonda.
51/53 Robert Street,
Northampton NN1 2NQ
01604 35937/30191

Pacet Manufacturing Limited
Electric fan replacement kits and Hot Start warm-up device, both designed to aid efficiency and improve mpg.
Wyebridge House, Cores End Road,
Bourne End, Bucks SL8 5HH
01628 526754

Royale Motor Company
Manufacturers of the Royale Drophead 1930s look-alike, using Jaguar components.
Unit 1 & 2 Pear Tree Street, Bamber Bridge, Preston, Lancs PR5 6EZ
01772 628462

SC Parts
High quality Falcon stainless steel exhaust systems for most XK applications.
13 Cobham Way, Gatwick Road,
Crawley, West Sussex RH10 2RX
01293 547841

SIP (Industrial Products Ltd)
Range of power and hand tools, pressure washers and Omega trolley jacks.
Gelders Hall Road, Shepshed,
Loughborough, Leics LE12 9NH
01509 503141

SU Carburettors
See Burlen.

Suffolk Sportscars
Manufacturers of the Suffolk SS100 Jaguar replica.
PO Box 55, Woolpit,
Bury St Edmunds, Suffolk IP30 9QS
01359 242277

Sykes-Pickavant
High quality, superbly finished hand tools for the Jaguar owner.
Warwick Works, Kilnhouse Lane,
Lytham St Annes, Lancs FY8 3DU
01253 721291

Techno-Weld
Low heat aluminium welding system.
Aston Works, Back lane, Aston,
Oxon OX18 2BX
01993 851028

Teng Tools
A large range of quality hand tools.
Unit 5, Flitwick Industrial Estate,
Maulden Road, Flitwick, Beds MK45 1UF
01525 718080

VSE Engineering
Jaguar XK engine specialist. Rebuilds, tuning, engine component sales, unleaded conversions, etc. Instrumental in the compilation of the engine chapter of this book.
Llanbister, Llandindrod Wells,
Powys LD1 6TL
01597 840308

Weber Concessionaires Ltd
Importers of Weber carburettors and spares.
Dolphin Road
Sunbury
Middlesex TW16 7HE
01932 788805

CLUBS AND ORGANISATIONS

Jaguar Car Club
Membership Secretary, 'Barbary',
Chobham Road, Horsell, Woking,
Surrey GU21 4AS
01483 763811

Jaguar Drivers Club
Jaguar House, 18 Stuart Street,
Luton, Beds LU1 2SL
01582 419332

Jaguar Enthusiasts' Club Ltd
Freepost, (BS-6705) Patchway,
Bristol BS12 6BR

SPECIALIST MAGAZINES

Jaguar World
Bi-monthly (at time of writing) magazine dealing solely with Jaguar matters.
Kelsey Publishing Ltd, Kelsey House,
77 High Street, Beckenham,
Kent BR3 1AN
0181 658 3531